Synthese Library

Studies in Epistemology, Logic, Methodology,
and Philosophy of Science

Volume 383

Editor-in-Chief
Otávio Bueno, University of Miami, Department of Philosophy, USA

Editors
Berit Brogaard, University of Miami, USA
Anjan Chakravartty, University of Notre Dame, USA
Steven French, University of Leeds, UK
Catarina Dutilh Novaes, University of Groningen, The Netherlands

More information about this series at http://www.springer.com/series/6607

David Atkinson • Jeanne Peijnenburg

Fading Foundations

Probability and the Regress Problem

David Atkinson
University of Groningen
Groningen, The Netherlands

Jeanne Peijnenburg
University of Groningen
Groningen, The Netherlands

Synthese Library
ISBN 978-3-319-58294-8   ISBN 978-3-319-58295-5 (eBook)
DOI 10.1007/978-3-319-58295-5

Library of Congress Control Number: 2017941185

© The Editor(s) (if applicable) and The Author(s) 2017. This book is an open access publication
Open Access This book is distributed under the terms of the Creative Commons Attribution 4.0 International License (http://creativecommons.org/licenses/by/4.0/), which permits use, duplication, adaptation, distribution and reproduction in any medium or format, as long as you give appropriate credit to the original author(s) and the source, provide a link to the Creative Commons license and indicate if changes were made.

The images or other third party material in this book are included in the work's Creative Commons license, unless indicated otherwise in the credit line; if such material is not included in the work's Creative Commons license and the respective action is not permitted by statutory regulation, users will need to obtain permission from the license holder to duplicate, adapt or reproduce the material.

The use of general descriptive names, registered names, trademarks, service marks, etc. in this publication does not imply, even in the absence of a specific statement, that such names are exempt from the relevant protective laws and regulations and therefore free for general use.

The publisher, the authors and the editors are safe to assume that the advice and information in this book are believed to be true and accurate at the date of publication. Neither the publisher nor the authors or the editors give a warranty, express or implied, with respect to the material contained herein or for any errors or omissions that may have been made. The publisher remains neutral with regard to jurisdictional claims in published maps and institutional affiliations.

Printed on acid-free paper

This Springer imprint is published by Springer Nature
The registered company is Springer International Publishing AG
The registered company address is: Gewerbestrasse 11, 6330 Cham, Switzerland

Tief ist der Brunnen der Vergangenheit. Sollte man ihn nicht unergründlich nennen? (Deep is the well of the past. Should one not call it unfathomable?)

Thomas Mann — *Joseph und seine Brüder*

Preface

This book is the result of ten years on and off thinking about infinite regresses in epistemology. It draws on several of our papers, which, partly because of the development of our thoughts, are not always well connected.

Our overall purpose here is to show how our understanding of infinite epistemic chains benefits from an analysis of justification in terms of probability theory. It has been often assumed that epistemic justification is probabilistic in character, but we think that the consequences of this assumption for the epistemic regress problem have been insufficiently taken into account.

The book has eight chapters, detailed calculations having been relegated to appendices. Chapter 1 contains an introduction to the epistemological regress problem, giving some historical background, and recalling its three attempted solutions, foundationalism, coherentism and infinitism. Chapter 2 discusses different views on epistemic justification, since they bear on both the framing of the problem and its proposed solution. Chapters 3 and 4 form the core of the book. Taking as our point of departure a debate between Clarence Irving Lewis and Hans Reichenbach, we introduce the concept of a probabilistic regress, and we explain how it leads to a phenomenon that we call fading foundations: the importance of a foundational proposition dwindles away as the epistemic chain lengthens. In Chapters 5 and 6 we describe how a probabilistic regress resists the traditional objections to infinite epistemic chains, and we reply to objections that have been raised against probabilistic regresses themselves. Chapter 7 compares a probabilistic regress to an endless hierarchy of probability statements about probability statements; it is demonstrated that the two are formally equivalent. In the final chapter we leave one-dimensional chains behind and turn to multi-dimensional networks. We show that what we have found for linear chains applies equally to networks that stretch out in many directions: the effect of foundational propositions fades away as the network expands.

Epistemic regresses are not the only regresses about which philosophers have wracked their brains. The ancient Greeks and the mediaeval scholastics worried a lot about infinite causal chains, and more recently philosopers have shown interest in the phenomenon of grounding. Although we remain silent about the latter, and only tangentially touch upon the former, we believe that our analysis could shed light on causal regresses — on condition that causality is interpreted probabilistically.

We owe much to others who have concurrently been thinking about epistemic regresses, notably Peter D. Klein and Scott F. Aikin. Peter Klein deserves the credit for being the first to set the cat among the pigeons by supposing that infinite regresses in epistemology are not *prima facie* absurd. With Scott Aikin one of us organized a workshop on infinite regresses in October 2013 at Vanderbilt University. This resulted in a special issue of *Metaphilosophy* (2014, vol. 45 no. 3), which was soon followed by a special issue of *Synthese* (2014, vol. 191 no. 4), co-edited with Sylvia Wenmackers.

The writing of this book has been made possible by financial support from the Dutch Organization for Scientific Research (Nederlandse Organisatie voor Wetenschappelijk Onderzoek, NWO), grant number 360-20-280. Our colleagues at the Faculty of Philosophy of the University of Groningen provided support of many different kinds. This has meant a lot to us and we thank them very much.

Aix-en-Provence, October 2015

David Atkinson and Jeanne Peijnenburg

Contents

1 The Regress Problem 1
   1.1 Reasons for Reasons: Agrippa's Trilemma 1
   1.2 Coherentism and Infinitism 8
   1.3 Vicious Versus Innocuous Regress 14

2 Epistemic Justification 25
   2.1 Making a Concept Clear 25
   2.2 Two Questions 29
   2.3 Entailment 33
   2.4 Probabilistic Support 36
   2.5 Smith's Normic Support 45
   2.6 Alston's Epistemic Probability 51

3 The Probabilistic Regress 59
   3.1 A New Twist 59
   3.2 The Lewis-Reichenbach Dispute 61
   3.3 Lewis's Argument 65
   3.4 A Counterexample 69
   3.5 A Nonuniform Probabilistic Regress 72
   3.6 Usual and Exceptional Classes 74
   3.7 Barbara Bacterium 78

4 Fading Foundations and the Emergence of Justification 83
   4.1 Fading Foundations 83
   4.2 Propositions versus Beliefs 86
   4.3 Emergence of Justification 90
   4.4 Where Does the Justification Come From? 94
   4.5 Tour d'horizon 98

5	**Finite Minds** 101	
	5.1 Ought-Implies-Can 101	
	5.2 Completion and Computation 105	
	5.3 Probabilistic Justification as a Trade-Off 107	
	5.4 Carl the Calculator 115	
6	**Conceptual Objections** 119	
	6.1 The No Starting Point Objection 119	
	6.2 A Probabilistic Regress Needs No Starting Point 124	
	6.3 The Reductio Argument 128	
	6.4 How the Probabilistic Regress Avoids the Reductio 131	
	6.5 Threshold and Closure Constraints 134	
	6.6 Symmetry and Nontransitivity 139	
7	**Higher-Order Probabilities** 143	
	7.1 Two Probabilistic Regresses 143	
	7.2 Second- and Higher-Order Probabilities 145	
	7.3 Rescher's Argument 150	
	7.4 The Two Regresses Are Isomorphic 156	
	7.5 Making Coins 160	
8	**Loops and Networks** 167	
	8.1 Tortoises and Serpents 167	
	8.2 One-Dimensional Loops 169	
	8.3 Multi-Dimensional Networks 173	
	8.4 The Mandelbrot Fractal 177	
	8.5 Mushrooming Out 182	
	8.6 Causal Graphs 185	
A	**The Rule of Total Probability** 191	
	A.1 Iterating the rule of total probability 192	
	A.2 Extrema of the finite series 193	
	A.3 Convergence of the infinite series 194	
	A.4 When does the remainder term vanish? 195	
	A.5 Example in the usual class 196	
	A.6 Example in the exceptional class 197	
	A.7 The regress of entailment 198	
	A.8 Markov condition and conjunctions 199	
B	**Closure Under Conjunction** 203	

C	**Washing Out of the Prior** 207	
	C.1 Washing out 207	
	C.2 Example: a bent coin 209	
	C.3 Washing out is not fading away 211	
D	**Fixed-Point Methods** 213	
	D.1 Linear iteration 213	
	D.2 Quadratic Iteration 214	

References 219

Index 233

Chapter 1
The Regress Problem

Abstract
The attempt to justify our beliefs leads to the regress problem. We briefly recount the problem's history and recall the two traditional solutions, foundationalism and coherentism, before turning to infinitism. According to infinitists, the regress problem is not a genuine difficulty, since infinite chains of reasons are not as troublesome as they may seem. A comparison with causal chains suggests that a proper assessment of infinitistic ideas requires that the concept of justification be made clear.

1.1 Reasons for Reasons: Agrippa's Trilemma

We believe many things: that the earth is a spheroid, that Queen Victoria reigned for more that sixty years, that Stockholm is the capital of Finland, that the Russians were the first to land on the moon. Some of these beliefs are true, others are false. A belief might be true by accident. Suppose I have a phobia which makes me believe that there is a poisonous snake under my bed. After many visits to a psychiatrist and intensive therapy I gradually try to convince myself that this belief stems from traumatic and suppressed childhood experiences. One fine day I finally reach the point where I, nervous and trembling, force myself to get into bed before first looking under it. Unbeknownst to me or the psychiatrist, however, a venomous snake has escaped from the zoo and has ensconced itself under my bed. My belief in the proposition 'There is a poisonous snake under my bed' is true, but it is accidentally true. I do not have a good reason for this belief, since I am ignorant of the escape and agree with the psychiatrist that reasons based on my phobia are not good reasons.

If however a belief is based on good reasons, we say that it is epistemically justified. Had I been aware of the fact that the snake had escaped and in fact had made its way to my bedroom, I would have been in possession of a good reason, and would have been epistemically justified in believing that the animal was lying under my bed.

According to a venerable philosophical tradition, a true and justified belief is a candidate for knowledge. One of the things that is needed in order for me to *know* that there is a snake under my bed is that the good reason I have for it (namely my belief that the reptile had slipped away and is hiding in my room) is itself justified. Without that condition, my reason might be itself a fabrication of my phobic mind, and thus ultimately fall short of being a *good* reason.

What would count as a good reason for believing that a snake has escaped and installed itself in my bedroom? Here is one: an anxious neighbour knocks on my door, agitatedly telling me about the escape. But how do I know that what the neighbour says is true? It seems I need a good reason for that as well. My friendly neighbour shows me a text message on his cellphone, just sent by the police, which contains the alarming news. That seems to be quite a good reason — although, how do I know that the police are well informed? I need a good reason for that as well. I call the head of police, who confirms the news, and says that he was apprised of it by the director of the zoo; I call the director, who tells me that the escape has been reported to her by the curator of the reptile house, and so on. True, my actions are somewhat curious, and they may well signal that a phobia for snakes is not the only mental affliction that plagues me. The point however is not a practical but a principled one. It is that a reason is only a good reason if it is backed up by another good reason, which in turn is backed up by still another other good reason, and so on. We thus arrive at a chain of reasons, where the proposition 'There is a dangerous snake under my bed' (the target proposition q) is justified by 'A neighbour knocks on my door and tells me that a snake has escaped' (reason A_1), which is justified by 'The police sent my neighbour a text message about the escape' (reason A_2), which is justified by A_3, and so on:

$$q \longleftarrow A_1 \longleftarrow A_2 \longleftarrow A_3 \longleftarrow A_4 \ldots \tag{1.1}$$

Such a justificatory chain, as we shall call it, gives rise to the regress problem. It places us in a position where we have to choose between two equally unattractive options: either the chain must be continued, for otherwise we

1.1 Reasons for Reasons: Agrippa's Trilemma

cannot be said to *know* the proposition q, or the chain must come to a stop, but then it seems we are not justified in claiming that we really can know q, since there is no reason for stopping. Laurence Bonjour called considerations relating to the regress problem "perhaps the most crucial in the entire theory of knowledge", and Robert Audi observes that no epistemologist quite knows how to handle the problem.[1]

The roots of the regress problem extend far back into epistemological history, and scholars often refer to the Greek philosopher Agrippa. Little is known about Agrippa, apart from the fact that he probably lived in the first century A.D. and might have been among the group of sceptics discussed by Sextus Empiricus, a philosopher and practising physician who allegedly flourished a century later. Sextus' most famous work, *Outlines of Pyrrhonism*, contains an explanation and defence of what he takes to be the philosophy of another shadowy figure, namely Pyrrho of Elis (*c.* 365–270 B.C.), who himself wrote nothing, but became known for his sober life style and his aversion to academic or theoretical reasoning. So-called Pyrrhonian scepticism advocates the attainment of *ataraxia*, a state of serene calmness in which one is free from moods or other disturbances. An important technique for reaching this state is the practicing of argument strategies known as *tropoi* or modes, i.e. means to engender suspension of judgement by undermining any claim that conclusive knowledge or justification has been attained. For example, if it were claimed that a particular sound is known to be soft, a Pyrrhonian would point out that to a dog it is loud, and that we cannot judge the loudness or softness independently of the hearer. Typically, a Pyrrhonian will try to thoroughly acquaint himself with the modes, so that reacting in accordance with them becomes as it were a second nature. In this manner he will be able to routinely refrain from assenting to any weighty proposition q or $\neg q$, and thus avoid getting caught up in one of those rigid intellectual positions that he loathes so much.

In Book 1 of *Outlines of Pyrrhonism*, Sextus discusses five modes which he attributes to "the more recent Sceptics" (to be distinguished from what he calls "the older Sceptics"), and which Diogenes Laertius in the third century would identify with "Agrippa and his school".[2] Of these five modes the

---

[1] Bonjour 1985, p.18; Audi 1998, 183–184. The thought is echoed by Michael Huemer when he writes that regress arguments "concern some of the most fundamental and important issues in all of human inquiry" (Huemer 2016, 16).

[2] Sextus Empiricus, *Outlines of Pyrrhonism*, Book I, 164; see p. 40 in the translation *Outlines of scepticism* by Julia Annas and Jonathan Barnes. Diogenes Laertius, *Lives of eminent philosophers*, Volume 2, Book 9, 88. We thank Tamer Nawar and an anonymous referee for guidance in matters of ancient philosophy.

three that are of especial interest are the Mode of Infinite Regress, the Mode of Hypothesis, and the Mode of Circularity or Reciprocation. Here is how Sextus explains them:

> In the mode deriving from infinite regress, we say that what is brought forward as a source of conviction for the matter proposed itself needs another source, which itself needs another, and so on *ad infinitum*, so that we have no point from which to begin to establish anything, and suspension of judgement follows. ...We have the mode from hypothesis when the Dogmatists, being thrown back *ad infinitum*, begin from something which they do not establish but claim to assume simply and without proof in virtue of a concession. The reciprocal mode occurs when what ought to be confirmatory of the object under investigation needs to be made convincing by the object under investigation; then, being unable to take either in order to establish the other, we suspend judgement about both.[3]

In other words, whenever a 'dogmatist' (as Sextus calls any philosopher who is not a Pyrrhonian sceptic) claims that he knows a proposition q, the Pyrrhonian sceptic will ask him what his reason is for q. After the dogmatist has given his answer, for example reason A_1, the sceptic will ask further: what is your reason for A_1? In the end it will become clear that the dogmatist has only three options open to him, jointly known as 'Agrippa's Trilemma':

1. He goes on giving reasons for reasons for reasons, without end.
2. He stops at a particular reason, claiming that this reason essentially justifies all the others that he has given.
3. He reasons in a circle, where his final reason is identical to his first.

In the first case the justificatory chain is infinitely long, in the second case it comes to a halt, and in the third case it forms a loop. The sceptic is quick to point out that none of these options can be accepted as a justification for q. The first option is impossible from a practical point of view, since we are ordinary human beings with a restricted lifespan. Moreover, even if we were to live forever, continuing to give reason after reason, we would never reach the origin of the justification, since by definition the chain does not have an origin. The second option is also unsatisfying. For why do we stop at this particular reason and not at another? If we can answer this question, we have a reason for what we claimed is without a reason, so we actually did not stop the chain. And if we cannot answer the question, then stopping at this particular reason is arbitrary. The third option is likewise unacceptable, for

---

[3] Sextus Empiricus, *Outlines of scepticism*. Book I, 166–169. Translation by Julia Annas and Jonathan Barnes, 41.

1.1 Reasons for Reasons: Agrippa's Trilemma

justifying the object under investigation by calling on that very object is not particularly convincing.

The Pyrrhonian takes the moral of this discouraging story to be that we are never justified in claiming that we know a proposition q. Proposition q might be true, it might be false, we simply have no way to know for sure. The only viable option open to us is to suspend judgement. Suspension of judgement (*epoche*) does not imply that we will be paralyzed; it does not mean that we cannot form any beliefs, are incapable of making decisions, or cannot perform actions on the basis of these decisions. Although we should desist from making a truth-claim, it is perfectly acceptable to abide by appearances, customs, and natural inclinations, and to act in accordance with them. Thus, to return to our snake example, it is altogether acceptable and even recommended to take your neighbour's word for it and proceed correspondingly — that will actually make you a better, and at any rate a more normal person than to engage in highly abstract reasoning. The fact that we must take recourse to suspension of judgement should therefore not sadden of demoralize us. Quite the contrary. We should welcome this fact and embrace it, since that will free us from the futile and fruitless attempt to arrive at knowledge, certainty, or justified beliefs, and bring us closer to *ataraxia*.

Pyrrhonian scepticism appears to have been quite a popular philosophical outlook in the first century A.D. However interest in it slowly waned in the second and third century, and by the fourth the movement had practically disappeared.

About the same time that the Pyrrhonian movement petered out, appreciation for the ideas of the recently rediscovered Aristotle (384–322 B.C.) was on the rise. It turns out that Aristotle had anticipated something like the Agrippan Trilemma in his *Posterior Analytics* and in his *Metaphysics*. Unlike the Pyrrhonians, however, he does not use the trilemma as a means for arguing that we can never know a proposition. In fact the opposite is true. Rather than arguing that none of the three possibilities in Agrippa's Trilemma produces justification, Aristotle gives short shrift to possibilities one and three, and claims it to be evident that the second possibility *is* a proper justificatory chain, and so *does* give us knowledge of some kind, be it practical, theoretical, or productive. Here is how Aristotle phrases his position in the *Posterior Analytics*, where 'understanding' refers to what we have called 'knowledge', and where 'demonstration' is used for 'justification':

> Now some think that because one must understand the primitives there is no understanding at all; others that there is, but that there are demonstrations of everything. Neither of these views is either true or necessary.

> For the one party, supposing that one cannot understand in another way, claim that we are led back *ad infinitum* on the ground that we would not understand what is posterior because of what is prior if there are no primitives; and they argue correctly, for it is impossible to go through infinitely many things. And if it comes to a stop and there are principles, they say that these are unknowable since there is no *demonstration* of them, which alone they say is understanding; but if one cannot know the primitives, neither can what depends on them be understood *simpliciter* or properly, but only on the suspicion that they are the case.
>
> The other party agrees about understanding; for it, they say, occurs only through demonstration. But they argue that nothing prevents there being demonstration of everything; for it is possible for the demonstration to come about in a circle and reciprocally.
>
> But *we* say that neither is all understanding demonstrative, but in the case of the immediates it is non-demonstrable — and that this is necessary is evident; for if it is necessary to understand the things which are prior and on which the demonstration depends, and it comes to a stop at some time, it is necessary for these immediates to be demonstrable. So as to that we argue thus; and we also say that there is not only understanding, but also some principle by which we become familiar with the definitions.[4]

A similar reasoning can be found in the *Metaphysics*:

> There are [people who demand] that a reason shall be given for everything; for they seek a starting-point, and they wish to get this by demonstration, while it is obvious from their actions that they have no conviction. But their mistake is what we have stated it to be; they seek a reason for that for which no reason can be given; for the starting-point of demonstration is not demonstration.[5]

This is not the place, nor do we have the competence to deal with historical details or with intricacies of translation from the Greek. Relevant for our purpose is the observation that the above passages of Aristotle herald the birth of what in contemporary epistemology became known as *foundationalism*. Foundationalism comes in various shapes and sizes, but its essence is an adherence to a foundation, be it a basic belief, a basic proposition, or even a basic experience. It thus can be described as joining Aristotle in embracing the second option of Agrippa's trilemma. Like Aristotle, foundationalists maintain that justified beliefs come in two kinds: the ones that do, and the ones that do not depend for their justification on other justified beliefs. It is not always clear what the nature of the latter kind is, but in most versions of

---

[4] Aristotle 1984a, *Posterior Analytics*, Book I, Chapter 3, 72b 5-24. Translation by Jonathan Barnes, 117.

[5] Aristotle 1984c, *Metaphysics*, Book IV, Chapter 6, 1011a 3-13. Translation by W.D. Ross, 1596.

1.1 Reasons for Reasons: Agrippa's Trilemma

foundationalism these justified beliefs are in some sense self-evident and so not in need of other beliefs for their justification.

During the Middle Ages foundationalism became the dominant school of thought concerning the structure of justification. Especially Thomas Aquinas (1225-1274), whose Aristotelian outlook so greatly influenced Western epistemology, contributed to the view that the Agrippan Trilemma could be resolved by a foundationalist response to the regress problem. In his *Commentary on Aristotle's Posterior Analytics*, Aquinas starts by defending the traditional view that knowledge (*scientia*) of a proposition q implies that one has a particular kind of justification for q. The justification for q is either inferential or non-inferential. In the first case q is justified by another proposition, for example A_1, that is both logically and epistemically prior to q; here we know q *per demonstrationem*, that is through A_1. In the second case we know q by virtue of itself (*per se nota*). Aquinas follows Aristotle in arguing that inferential justification cannot exist without non-inferential justification. We may know many propositions *per demonstrationem*, but in the end every justificatory chain must culminate in a proposition that we know *per se*.

The end of the fifteenth century evinced renewed interest in Sextus Empiricus, whose texts were brought to Italy from Byzantium. A Latin translation of Sextus' *Outlines*, which appeared in 1562 in Paris under the title *Pyrrhoniarum Hypotyposes*, kindled the interest of European humanists, who had a taste for using sceptical arguments in their attack not only on astrology and other pseudo-science, but also on mediaeval scholasticism and forms of all too rigid Aristotelianism.[6] An important rôle in the revival of Pyrrhonian scepticism in the sixteenth century was played by the French philosopher and essayist Michel de Montaigne (1533–1592). In the manner of Sextus and Pyrrho, Montaigne stressed that knowledge cannot be obtained, and that we should suspend judgement on all matters. He accordingly propagated tolerance in moral and religious matters, as Pyrrho had done, and espoused an undogmatic adoption of customs and social rules.

Although Montaigne's work was highly influential at the beginning of the seventeenth century, his impact was soon overshadowed by the authority of his compatriot René Descartes (1596–1650). This supersession turned out to be definitive: when today epistemologists talk about philosophical scepticism, they generally have Descartes rather than Montaigne or Pyrrho in mind. Cartesian scepticism is however quite different from scepticism in the

---

[6] Thanks to Lodi Nauta for helpful conversations.

Pyrrhonian vein.[7] Whereas Pyrrhonians cheerfully embrace the adage that knowledge cannot be had because information obtained by the senses and by reason is unreliable, Descartes aims at no less than a theory of everything, a coherent framework that could explain the entire universe. The way in which he tried to reach this goal has become part of the canon: in an attempt to arrive at a proposition that can resist all doubt, so as to make it the basis on which to erect his all encompassing framework, Descartes applies his sceptical method of doubting every proposition that could possibly be false. Thus he arrives at the allegedly indubitable truth of the *cogito ergo sum*. But of course, the adherence to the *cogito* as the foundation for all our knowledge eventually makes him more a foundationalist than a sceptic. In a sense, the difference between the two kinds of scepticism could not be greater: whereas a Pyrrhonian uses the sceptical method as a means towards *ataraxia*, the state of imperturbability where one is at peace with the supposed fact that knowledge cannot be had, for Descartes it is a way of acquiring knowledge of the entire external world and of our place therein.

1.2 Coherentism and Infinitism

Already in the seventeenth century there was severe criticism of the *cogito*, and of the whole Cartesian method of doubt. The foundationalist thrust of Descartes' philosophy, however, was generally accepted, since it harmonized perfectly with the dominant tradition in epistemology. Most philosophers before Descartes were foundationalists concerning justification, as were many after him. The English empiricists of the eighteenth century, John Locke, George Berkeley, and David Hume, all had a foundationalist outlook. The same can in a sense be said of the great German philosopher of the Enlightenment, Immanuel Kant, although he appears to have been a bit more cautious. In his *Critique of Pure Reason* he emphasizes that from the fact that every event has a cause, it does not follow that there is a cause for everything. Similarly, from the fact that every proposition has a reason, it does not follow that there is a reason for the entire justificatory chain.[8] Yet, says

---

[7] For a good explanation of the differences between Cartesian and Pyrrhonian scepticism, see Williams 2010.

[8] The difference is nowadays known as one of the scope distinctions. The statement 'For each y there is an x to which y stands in the relation R' ($\forall y\ \exists x\ yRx$) differs from 'There is an x to which each y stands in the relation R' ($\exists x\ \forall y\ yRx$). Standard example: 'Every mammal has a mother' differs from 'There is something that is the

1.2 Coherentism and Infinitism

Kant, humans have a natural inclination to posit such a foundational cause or reason, and Kant's text does not always make it very clear whether this inclination should be resisted or put to practical use.

In the nineteenth century, Hegel developed an anti-foundationalist epistemology, as did Nietzsche, but it was not until the twentieth century that a serious alternative to foundationalism surfaced in the form of coherentism (although major figures in the twentieth century like Bertrand Russell, Alfred Ayer, and Rudolf Carnap remained convinced foundationalists). The main motivation behind the rise of coherentism was dissatisfaction with the foundationalist approach, especially with the idea that basic beliefs are somehow self-justifying and could exist autonomously. "No sentence enjoys the *noli me tangere* which Carnap ordains for the protocol sentences", writes Otto Neurath in 1933 about Carnap's attempt to logically re-erect the world from a bedrock of basic elements or protocol sentences, as he calls them.[9] According to Neurath and other coherentists, sentences are always compared to other sentences, not to experiences or 'the world' or to sentences that have some sort of sovereign standing.[10]

Coherentism is described in many textbooks as the attempt to put an end to the regress problem by embracing the third alternative of Agrippa's Trilemma. For example, A_2 can be a reason for A_1, which is a reason for q, which in turn is a reason for A_2. The position is however markedly more sophisticated: rather than advocating reasoning in a circle, it maintains that justification is not confined to a finite or ring-shaped justificatory chain. What is justified, according to coherentists, are first and foremost entire systems of beliefs or propositions, not individual elements in these systems. Justification of individual beliefs through one-dimensional justificatory loops is a special case only, a degenerate form of the holistic process that constitutes justification.

According to coherentism, the more coherent a system is, the more it is justified. But what exactly does it mean to say that beliefs in a system cohere with one another in that system? Twentieth century coherentists have worked hard to find a satisfying definition of 'coherence', but Laurence Bonjour has argued that there is no simple answer to the question, since coherence de-

---

mother of all mammals'. The difference was already acknowledged in the Middle Ages and perhaps even by Aristotle, but has not always been applied consistently across the board.

[9] Neurath 1932-1933, 203. See also Carnap 1928.

[10] In the telling words of Donald Davidson: "what distinguishes a coherence theory is simply the claim that nothing can count as a reason for holding a belief except another belief" (Davidson 1986, 310).

pends on many different conditions being fulfilled; in fact, an entire list of different coherence criteria can be made.[11] A complicating factor in finding a definition of coherence is that we want this definition also to incorporate a *measure*, so that we can determine *how* coherent a particular system is. Many ingenious suggestions for formal coherence measures have been put forward.[12] All these measures are vulnerable to a classic criticism, namely that coherence is not truth-conducive: a system of propositions can be coherent to the highest degree while all of the propositions are in fact false. The criticism was already ventilated by Bertrand Russell at the beginning of the twentieth century and is sometimes referred to as the Bishop Stubbs objection:

> Whatever the standards of coherence may be, it seems likely that alternative sets of propositions will meet them: as Russell 1906 pointed out, although the highly respectable Bishop Stubbs died in his bed, the proposition "Bishop Stubbs was hanged for murder" can readily be conjoined with a whole group of others to form a set which passes any plausible coherence test; and indeed, the same can be said of the propositions that make up any good work of realistic fiction.[13]

In fact the Bishop Stubbs objection to coherentism cuts even deeper than Russell envisaged. As Luc Bovens and Stephan Hartmann showed in 2003, a system which is more coherent than another system cannot even be said to have a *higher probability* of being true than the other system.[14]

At the beginning of the twenty-first century a third approach to the epistemological regress problem entered the philosophical arena, one that is now known as 'infinitism'. While foundationalism and coherentism are said to avoid the regress problem by opting for the second, respectively the third, possibility of Agrippa's Trilemma, infinitism chooses the first. According to infinitists, it is not *prima facie* absurd that the process of giving reasons for reasons might go on without end, so that the justificatory chain will be infinitely long.

---

[11] Bonjour 1985, 97-99.
[12] See for example Olsson 2001, 2002, 2005a, 2005b; Shogenji 1999. For the relation between coherence and confirmation, see Fitelson 2003; Dietrich and Moretti 2005; Moretti 2007. For defences of coherentism in general, see Quine and Ullian 1970; Rescher 1973; Bonjour 1985; Davidson 1986; Lehrer 1997.
[13] Walker 1997, 310. Although in this quotation Walker refers to propositions, a similar objection, albeit one that is somewhat more complicated, could be made with reference to beliefs. Ibid. 316. See for Russell's argument, Russell 1906.
[14] Bovens and Hartmann 2003. See also Olsson 2005b.

1.2 Coherentism and Infinitism

To say that infinitism has consistently had a bad press would be claiming too much. Infinitism had no press at all, since until very recently nobody took it seriously. The reason for this is not difficult to discern. In an epistemological tradition dominated by Aristotelian and Cartesian foundationalism, a position like infinitism is highly counterintuitive to say the least; for how could anybody, in Aristotle's words, "go through infinitely many things"? It is therefore not surprising that infinitism is hardly, if ever, mentioned in treatises or textbooks; and if it is mentioned, then it usually serves as an example of a blatantly ridiculous way to go. Yet it cannot be denied that infinitism sits well with some modern ideas about the nature of knowledge, such as that knowledge is essentially fallible, and that the human search for it is, indeed, without end. Despite many attempts to show the contrary, it is not at all clear how these ideas, which so many of us endorse, can be smoothly combined with foundationalism or even coherentism.[15]

In this book we will investigate the consequences of an infinitist response to the regress problem. We do not propose to defend infinitism as such. Rather our aim is twofold. On the one hand, we intend to show that some standard objections to the position are not as strong as they might seem at first sight. On the other hand, we explain how our analysis of these objections brings about insights that cast new light on the traditional positions, foundationalism and coherentism; as we will see, a careful analysis of infinite justificatory chains will teach us interesting novel facts about finite ones. In the end we somehow try to get it all, sketching the contours of an infinitist version of coherentism, which also acknowledges the foundationalist lesson that we should somehow make contact with the world. We will return to this in the final chapter.

All-important for the development of infinitism was the work by Peter Klein. Around 2000 Klein wrote a number of papers in which he took the bull by the horns and presented infinitism as a genuine competitor to coherentism and foundationalism. Here is how Klein introduces his view in a relatively early paper:

> The purpose of this paper is to ask you to consider an account of justification that has largely been ignored in epistemology. When it has been considered, it has usually been dismissed as so obviously wrong that arguments against it are not necessary. The view that I ask you to consider can be called "Infinitism". Its central thesis is that the structure of justificatory reasons is infinite and nonrepeating. My primary reason for recommending infinitism is that it can

---

[15] For a prominent attempt at reconciling foundationalism and fallibilism, see Audi 1998.

provide an acceptable account of *rational beliefs*, i.e. beliefs held on the basis of adequate reasons, while the two alternative views, foundationalism and coherentism, cannot provide such an account."[16]

Klein is a convinced advocate of infinitism. As he sees it, infinitism is not just a third way to solve the regress problem beside two other approaches — it is the *only* viable solution to the regress problem.[17] The reason is that infinitism is the only account that can satisfy "two intuitively plausible constraints on good reasoning" which jointly entail that the justificatory chain is infinite and non-repeating.[18] The two constraints are the Principle of Avoiding Circularity (PAC) and the Principle of Avoiding Arbitrariness (PAA). Here is Klein about the first constraint:

PAC: For all q, if a person, S, has a justification for q, then for all A_i, if A_i is in the evidential ancestry of q for S, then q is not in the evidential ancestry of A_i for S.[19]

By the term 'evidential ancestry' Klein refers to the order of the links in the justificatory chain for q. So in our justificatory chain (1.1), proposition A_2 is in the evidential ancestry of A_1 and q, and A_3 is in the evidential ancestry of A_2, A_1 and q. Klein considers PAC to be "readily understandable and requires no discussion", and hence refrains from further defending it.[20]

---

[16] Klein 1999, 297. The term 'infinitism' was however not coined by Klein. He gives the credits for inventing the term to Paul Moser, who uses "epistemic infinitism" to refer to "inferential justification via infinite justificatory regresses" (Moser 1984, 199). See Klein 1998, 919, footnote 1. Charles Sanders Peirce is often paraded as the first infinitist (Peirce 1868), but James Van Cleve has suggested that what Peirce actually defends is "the possibility that each cognition of an object be 'determined' by an earlier cognition", not the possibility of an infinite regress of *justification* (Van Cleve 1992, 357, footnote 29).

[17] "I conclude that neither foundationalism nor coherentism provides an adequate non-skeptical response to the epistemic regress problem. Only infinitism does." (Klein 2011a, 255); "...only infinitism is left as a possible solution on offer to the regress problem" (Klein 2007, 16). In his later work, however, Klein made a plea for a "rapprochement" between foundationalism and infinitism by arguing that basic beliefs are contextual: whether a particular belief is basic or not depends on the context (Klein 2014). John Turri also made an attempt to bring foundationalism and infinitism together by presenting an exampe of a justificatory chain which, although infinite, can nevertheless be handled by foundationalists (Turri 2009, 161-163). For criticism of this example, see Peijnenburg and Atkinson 2011, Section 6, and Rescorla 2014, 181-182.

[18] Klein 1999, 298.

[19] Ibid., 298-299. For convenience we have adjusted Klein's notation.

[20] Klein 2005, 136.

1.2 Coherentism and Infinitism

The Principle of Avoiding Arbitrariness is:

PAA: For all q, if a person, S, has a justification for q, then there is some reason, A_1, available to S for q; and there is some reason, A_2, available to S for A_1; etc.

In contrast to the first constraint, PAA is likely to generate a lot of discussion. For what does it mean to say that a proposition A_n is *available* to S as a reason for A_{n-1}? The answer to this question is clearly very important, for it involves what we mean by 'epistemic justification', and thus what we mean by the arrow in our justificatory chain:

$$q \longleftarrow A_1 \longleftarrow A_2 \longleftarrow A_3 \longleftarrow A_4 \ldots$$

Although Klein acknowledges the importance of the question, he believes that the discussion about the pros and cons of infinitism can be carried out without delving into the matter. He argues that A_n is available to S as a reason for A_{n-1} if and only if A_n is both *objectively* and *subjectively* available. *Objective availability* is about the relation between two propositions: A_n is objectively available as a reason for A_{n-1} if and only if it really *is* a reason for A_{n-1}. Klein remarks that what makes a proposition a reason "need not be fleshed out", since "there are many alternative accounts that could be employed by the infinitist"; hence the "thorny issue" of what makes a proposition a reason "can be set aside".[21] *Subjective availability* is about the relation between a proposition and a person: A_n is subjectively available as a reason to S if and only if A_n is "appropriately 'hooked up' to S's beliefs and other mental contents".[22] It need not imply that S actually believes or endorses A_n; it only means that S must in some sense be able to "call on" A_n.[23] For example, it is not necessary for S to know or believe that $366 + 71 = 437$ in the sense in which S knows or believes that $2 + 2 = 4$. It is enough for subjective availability if S is able to do the calculation when called on to do so. In Klein's words, "The proposition that $366 + 71 = 437$ is subjectively available to me because it is correctly hooked up to already formed beliefs."[24]

Unlike Klein, we do not believe that an investigation into the viability of infinitism can evade the question as to what makes a proposition a reason for

---

[21] Ibid., 136-137.

[22] Ibid., 136.

[23] Klein 1999, 300, 308-309.

[24] Ibid., 308. Coos Engelsma has argued that Klein's distinction between objectively and subjectively available can be variously interpreted (Engelsma 2015, Engelsma 2014).

another proposition. Thorny as the issue may be, the meaning of 'justification' cannot be set aside if we want to examine whether chains of justification must be finite or can be infinite. Klein is right that there exist many different accounts of epistemic justification, but it is not so that all these accounts can be used without problem. Some of the accounts will be useful to infinitists, while others might be more advantageous to foundationalists or coherentists. It is therefore important to have an account of justification, however provisional it may be, on which everybody agrees, and then see whether this account allows infinite justificatory chains — and if so, in what sense.

William Alston has argued that such a neutral account of justification is impossible.[25] In his view, no definition of justification can serve as an impartial starting point or as a tool for adjudicating epistemological debates. Every definition will eventually take sides, and favour a particular position in the epistemological debate about the structure of justfication. Alston's advice to the epistemological community therefore is to abstain from attempts at defining justification and instead turn to spelling out what he calls 'epistemic desiderata'. That will be more fruitful for the theory of knowledge than undertaking ill-fated attempts to find a definition of justification.

Alston's point is well taken, but we think it applies primarily to material accounts of justification, less so to formal ones. As we will argue in Chapter 2, focusing on the formal properties of epistemic justification might generate more consensus than Alston deems possible. Moreover, as we will show in Chapters 3 to 6, a focus on formal properties casts doubt on several objections to the idea that justificatory chains can be infinitely long. In the end, our formal explication of justification provides us with means to preserve many, although not all, of Peter Klein's intuitions about the value of infinitism.

1.3 Vicious Versus Innocuous Regress

Epistemology is of course not the only place where infinite regresses occur. They can also be found in other philosophical disciplines, as well as in areas outside philosophy. Many of these regresses are not troublesome at all. Especially mathematics abounds with regresses that are benign: every integer has both a successor and a predecessor, every line segment can be divided into two, every natural number can be doubled, and so on. Outside mathematics there are benign regresses too, such as the regress arising from the statement

---

[25] Alston 1989, 1993, 2005a.

1.3 Vicious Versus Innocuous Regress

that, in arriving at the Louvre, I had already reached the midpoint of the distance to the Louvre, and the midpoint of the distance to the first midpoint, and so on.

How do we distinguish between between vicious and harmless regresses? This is an intriguing question, but an attempt to answer it might be overly ambitious. As Daniel Nolan has argued, it will be difficult if not impossible to find a general answer: there simply is not one criterion that applies to all cases.[26] A more feasible plan, although still not an easy one, is to ask ourselves why exactly it is that *justificatory* regresses are widely perceived as vicious. Why are infinite justificatory chains readily consigned to the bad batch? The fact *that* they have been treated with hostility or neglect goes almost without saying. "It can hardly be pretended", writes David Armstrong, "that this reaction to the regress [i.e. calling it virtuous] has much plausibility. ... it is a *desperate* solution, to be considered only if all others are clearly seen to be unsatisfactory".[27] Here are a few more quotations that serve as illustrations. All are taken from epistemology textbooks which were published after Peter Klein launched his controversial view, for earlier books are often simply silent about the possibility.

> We humans, for better or worse, do not have an infinite amount of time. ... Evidently, then, proponents of infinitism have some difficult explaining to

---

[26] Nolan 2001. The same point is made by Nicholas Rescher (2010): "There is nothing vicious about regresses as such" (ibid., 21); "Infinite regression is not something that is absurd as such, involving by its very nature a fault or failing that can be condemned across the board. Its viciousness will depend on the specifics of the case." (ibid., 62). Even so, Rescher offers several rules of thumb for distinguishing a benign from a vicious regress. One of them involves the difference between regresses that are time-compressible and those that are not: the former are often harmless, but the latter may well be vicious: "any regress that requires the realization of an infinitude of [not time-compressible] actions is thereby vicious" (ibid., 53). A related distinction is that between consequences or co-conditions on the one hand and preconditions or pre-requisites on the other hand (ibid., 55-61). The former are time-compressible, the latter are not, so a regress with consequences or co-conditions will often be harmless while a regress with pre-conditions or pre-requisites will mostly be vicious. We briefly return to time-compressibility in Chapter 5.

Michael Huemer has made the interesting suggestion that an infinite regress is vicious (i.e. cannot exist) if it requires the instantiation of "an infinite intensive magnitude" (Huemer 2014, 88). He considers this suggestion to be a first step towards "a new theory of the vicious infinite" (ibid., 95). On evaluating infinite regress arguments in general, see Gratton 2009, which is a study in argumentation theory; Wieland 2014 also deals with the subject.

[27] Armstrong 1973, 155.

do. As a result, infinitism has attracted very few public supporters throughout the history of epistemology. It is, nonetheless, a logically possible approach to the regress problem, at least according to some philosophers.[28]

> The least plausible ... response to Agrippa's trilemma involves ... holding that an infinite chain of justification *can* justify a belief. The position is known as infinitism. On the face of it, the view is unsustainable because it is unclear how an infinite chain of grounds could ever justify a belief any more than an infinite series of foundations could ever support a house. Nevertheless, this view does have some defenders ...[29]

> [Infinitism] tells us that evidential chains can be infinitely long, and so need not terminate. [It] allows that [a belief] can be supported by an evidential chain that has an infinite number of links ... Such an infinite chain would have no final or terminating link. One difficulty with this option is that it seems psychologically impossible for us to have an infinite number of beliefs. If it is psychologically impossible to have an infinite number of beliefs, then none of our beliefs can be supported by an infinite evidential chain.[30]

> For one thing, justifications that never come to an end are not the sort of justifications we typically prize from the standpoint of learning more about the world. For another, [infinitism] seemingly would commit us to the idea that humans have an infinite chain of beliefs. ... Although the normal person undoubtedly has an indefinitey large number of beliefs, that person is unlikely to have a limitless supply of beliefs.[31]

Note that three of the four cited authors criticize infinitism because it supposedly implies that people have an infinite number of beliefs. The complaint dates back as far as Aristotle, and is known as the finite mind objection. We discuss this objection in Chapter 5. For the moment we restrict ourselves to observing that the intuition behind the finite mind objection is not so natural and widely shared as it may seem at first sight. Even among philosophers opposed to infinitism, there are some who *do* believe that people can have an infinite number of beliefs. Richard Fumerton, for example, writes in his paper on classical foundationalism:

---

[28] Moser, Mulder, and Trout 1998, 82.
[29] Pritchard 2006, 36.
[30] Lemos 2007, 48.
[31] Crumley 2009, 109-110.

1.3 Vicious Versus Innocuous Regress

> Klein is right that we do have an infinite number of beliefs, but I think he misses the real point of the regress argument for noninferentially justified beliefs. The viciousness of the regress is, I believe, *conceptual*.[32]

Do we have a finite mind? This is not so clear. We have finite brains, and minds supervene on brains, but does that mean that our mind is finite? What exactly does it mean to have a finite mind? That we cannot have an infinite number of beliefs? But how to count? Moreover, even if we have a finite mind in the sense that our beliefs are finite and therefore countable, this does not prevent us from saying many cogent things about infinities — how is that possible?

The routine manner in which epistemologists have rejected infinite justificatory chains is reminiscent of the customary ways in which infinite *causal* chains have been cast aside. Again, Aristotle appears to have played a major rôle here. His familiar arguments against infinite causal chains in his *Physics* and *Metaphysics* became a well-entrenched part of the philosophical canon. Yet Aquinas and other mediaeval scholars had already pointed out that Aristotle's arguments may be more restricted than they appear: not every causal regress seems to be vicious, it all depends on what is meant by 'causal connection'. So let us take a closer look at Aristotle's objection to causal regresses and the criticism thereof by the mediaeval schoolmen. This might help us to see why exactly it is that justificatory regresses have been rejected without much ado, and to assess whether such a hasty rejection is appropriate. In Chapter 8, in the final section, we will discuss causal chains in a more modern setting, namely that of causal graphs.

Aristotle's main argument against a causal regress is that it purports to explain a phenomenon, but in fact fails to do so. Suppose an event, an object, or a process A is explained by saying that it is caused by B, and B is causally explained by pointing to C, and so on. If this series were to go on indefinitely, it would remain unclear why A occurred in the first place. The only way to explain the occurrence of A is to refer to a principal or first cause, i.e. something that causes all the other elements in the series, but is itself uncaused. Aristotle stresses that his argument is not confined to a particular kind of causation, but applies to any of the four different causes that he distinguishes, i.e. material, efficient, final or formal:

> Evidently there is a first principle, and the causes of things are neither an infinite series nor infinitely various in kind. For, on the one hand, one thing cannot proceed from another, as from matter, *ad infinitum* ... nor on the other hand

---

[32] Fumerton 2001, 7. We will say more about the conceptual objections to infinitism in Chapter 6.

can the efficient causes form an endless series ... Similarly the final causes cannot go on *ad infinitum*. ... And the case of the formal cause is similar. ... It makes no difference whether there is one intermediate or more, nor whether they are infinite or finite in number. But of series which are infinite in this way, and of the infinite in general, all the parts down to that now present are like intermediates; so that if there is no first there is no cause at all.[33]

Aristotle's argument is the most intuitive when he talks about causation as setting something in motion. Suppose object A moves because it is moved by object B, and B moves because it is moved by C, and so on. Then, unless the series comes to rest in an Unmoved Mover, we cannot explain why A moved in the first place:

Now this [a thing being in motion] may come about in either of two ways, either ... because of something else which moves the mover, or because of the mover itself. Further, in the latter case, either the mover immediately precedes the last thing in the series, or there may be one or more immediate links: e.g. the stick moves the stone and is moved by the hand, which again is moved by the man; in the man, however, we have reached a mover that is not so in virtue of being moved by something else. Now we say that the thing is moved both by the last and by the first of the movers, but more strictly by the first, since the first moves the last, whereas the last does not move the first, and the first will move the thing without the last, but the last will not move it without the first: e.g. the stick will not move anything unless it is itself moved by the man. If then everything that is in motion must be moved by something, and by something either moved by something else or not, and in the former case there must be some first mover that is not itself moved by anything else, while in the case of the first mover being of this kind there is no need of another (for it is impossible that there should be an infinite series of movers, each of which is itself moved by something else, since in an infinite series there is no first term) — if then everything that is in motion is moved by something, and the first mover is moved not by anything else, it must be moved by itself.[34]

In other words, if a man moves a stone by moving a stick, the movement of the stone is not explained by referring merely to the movement of the stick. We must point to the man who moves the stick, for without him the stick would be at rest. The man's own movement, however, cannot be explained in this manner, since the man is not moved by anybody or anything outside him — he moves himself.

---

[33] Aristotle 1984c, *Metaphysics*, Book II, Chapter 2, 994a, 1-19. Translation by W.D. Ross, 1570.

[34] Aristotle 1984b, *Physics*, Book VIII, Chapter 5, 256a, 4-21. Translation by R.P. Hardie and R.K. Gaye, 427-428.

1.3 Vicious Versus Innocuous Regress

Thomas Aquinas pointed out that Aristotle's picture of a causal regress appears to be too simple. There are at least *two* different causal regresses, each of them covering Aristotle's four causes, one being vicious and one being benign. Aquinas and other scholastics refer to the distinction as a causal series *per se* versus a causal series *per accidens*. The difference should not be confused with the distinction we mentioned in Section 1.1 between knowing a proposition *per se* and knowing it *per demonstrationem*. Nor should it be simply put on a par with the distinction between necessary and accidental properties. Causal series *per accidens* and *per se* are about the ways in which its members are ordered, i.e. the way in which the causes in the series are linked. A particular cause can have necessary properties but be linked to other causes in an accidental way. Conversely, a cause may have accidental properties, but be part of a series of which the members are ordered in an essential way.

In a causal series *per se* each intermediate member (that is each member except the first and the last) exerts causal power on its successor by virtue of the causal power exerted on this member by its predecessor. Aristotle's stone-stick-man example in the above citation involves such an essential ordering of causes. The stick causes the stone to move by virtue of the fact that the man causes the stick to move. This series consists of three elements, of which only the second (the stick) exerts causal power on its successor (the stone) by virtue of the causal power exerted on it by its predecessor (the man). Of course there will be more intermediate members if the essential ordering is longer. If for example the stone were to move a pebble, the stone would cause the pebble to move by virtue of the fact that it was moved by the stick. The salient point is that the intermediate members depend for their causing on their being caused.

Things are different in a causal series *per accidens*. Here each member (except the last) exerts power on its successor, but not by virtue of the causal power exerted on it by its predecessor. The standard example is Jacob, who was begotten by Isaac, who in turn was begotten by Abraham. Again we have a series of three elements, but none of them, not even the second one, causes by virtue of the fact that it is caused. Isaac fathers Jacob not because of the fact that he was fathered by Abraham, but because of having had intercourse with Rebecca. A stick needs a hand to move the stone, but Isaac does not need Abraham to sleep with Rebecca. Of course, Isaac needs Abraham for his existence: if Abraham had not existed, then Isaac would not have existed either. But neither Abraham nor Abraham's intercourse with Sarah is the cause of Isaac begetting Jacob. As Patterson Brown formulates it in his outstanding paper on infinite causal regressions:

Abraham's copulation causes Isaac's conception, Isaac's copulation causes Jacob's conception, Jacob's copulation causes Joseph's conception. Each member has one attribute qua effect (being conceived) and quite another attribute qua cause (copulating).[35]

In an essential ordering of causes, on the other hand, the attributes qua effect and qua cause coincide:

> ...it is the same function of the stick (namely, its locomotion) which both is caused by the movement of the hand and causes the movement of the stone. Again, a series where the fire heats the pot and the pot in turn heats the stew, causing it to boil, is also essentially ordered; for the warmth of the pot is both caused by the warmth of the fire and cause of the warmth of the stew, while the warmth of the stew is both caused by the warmth of the pot and cause of the stew's boiling.[36]

The above examples suggest that the causal relation in an essentially ordered series is *transitive*, whereas the causal relation in an accidentally ordered series is *intransitive*.[37] If the man moves the stick, and the stick moves the stone, then the man moves the stone. But if Abraham begets Isaac, and Isaac begets Jacob, then it is not the case that Abraham begets Jacob.

The scholastics all agree that an essential ordering of causes needs a first member, whereas an accidental ordering does not. Consider again the case where we explain the moving of object A by pointing to B. The idea here is that we have not really explained the movement of A if B is moved by C; at best we have only postponed the explanation of A's movement, or better: we have now dressed it up as the question of how to explain B's movement. Unless we arrive at a first mover X, embodying the origin of the movement, the cry for an explanation will not be deadened and the explanation of A's movement will be woefully incomplete.[38] The situation is entirely different in an accidental ordering of causes. If we explain Jacob's conception by referring

---

[35] Brown 1966, 517.

[36] Ibid.

[37] Ibid. R.G. Wengert tried to formalize the transitivity of essentially ordered causes by means of Gottlob's Frege's ancestral relation (Wengert 1971).

[38] C.J.F. Williams argued that Thomas Aquinas in his *Summa Theologiae* commits a *petitio principii*: by assuming that the only 'movers' are either first or second movers, Thomas excludes by fiat the possibility that an infinite sequence may be doing the moving (Williams 1960). J. Owens doubts whether Williams' critique "come[s] to grips with the argument of Aquinas in the argument's own medieval setting", but he grants the point "as it stands from any concrete background and time" (Owens 1962, 244).

1.3 Vicious Versus Innocuous Regress

to the fact that Isaac made love to Rebecca, we have given a full and satisfactory explanation. Of course, we could go further and ask for an explanation of Isaac's lovemaking. But such an explanation, as Patterson Brown deftly notes, "will center on his actions with Rebecca, rather than on his having been sired by Abraham."[39] Therefore, according to Thomistic schoolmen, a causal regress *per se* is vicious because an essential causal ordering needs a first member; but a causal regress *per accidens* is harmless since an accidental ordering can exist without a member that is the first. Aristotle, when talking about causality, seems however to have had in mind solely causal orderings and regresses *per se*.

Still, it is not at all easy to find out how exactly an essential causal ordering differs from an accidental one. Is it because the former is transitive and the latter intransitive? That seems unlikely, for one can think of accidental causal orderings that are transitive. For example, Abraham is an ancestor of Isaac, and Isaac of Jacob, but Abraham is also an ancestor of Jacob. This ordering is transitive, but it is not essential: it is not the case that Isaac's being an ancestor is caused by Abraham's being an ancestor or that it causes Jacob's being an ancestor. So while it is true that the Abraham-begetting-Isaac example is intransitive, the intransitivity might be a feature of the example, not of the fact that it illustrates an accidental causal ordering. Conversely, as Brown notes, the relation '*A* is moved by *B*' need not always be used in a transitive manner.[40]

Another difficulty, not less serious, concerns the question why exactly the mediaeval schoolmen thought that an essential causal regress is vicious and an accidental causal regress is harmless. Why is it that a causal ordering *per se* needs a first member and a causal ordering *per accidens* does not? Brown discusses the possibility that it is simultaneity that does the trick. The idea is that, because causes in an essential ordering occur simultaneously (the man, the stick, and the stone all moving at the same time), it is impossible to have an infinite number of causes. For were we to allow an infinity of causes all happening instantaneously, we would defy Aristotle's ban on actual infinities, and no true Aristotelian would ever go that far. In an accidental causal series, however, the causes are ordered chronologically and thus do not occur at the same time; if they were to be infinite in number, they would form a potential, not an actual infinity. However, Brown argues that it is not the supposed simultaneity which requires that an essentially ordered series has a first term. His argument is strong: Aristotle and his followers themselves

---

[39] Brown 1966, 523.
[40] Ibid., 518.

explicitly deny that the argument for a first cause is related to the question whether an infinite number of concurrent intermediate causes is possible or not.[41]

What does the above excursion concerning causal regresses teach us about justificatory regresses? We have said that the hostility towards justificatory regresses, early and late, parallels a hostility towards causal regresses, especially in Aristotle's work. However, we have seen that it makes sense to distinguish between two different causal regresses (even if the distinction is not always crystal clear and even if it is unclear whether the dichotomy is exhaustive). Thus the question arises whether the same goes for justficatory regresses. Can they be divided in a similar dichotomy? Is a typical justificatory regress more like a causal regress *per se* or is it more like a regress *per accidens*? Does it resemble the vicious man-stick-stone example or is it similar to the benign Abraham-begets-Isaac paradigm? With respect to all these questions, the jury is still out. Some philosophers apparently have the intuition that justificatory regresses mirror the man-stick-stone example, transitivity and all:

> Consider a train of infinite length, in which each carriage moves because the one in front of it moves. Even supposing that that fact is an adequate explanation for the motion of each carriage, one is tempted to say, in the absence of a locomotive, that one still has no explanation for the motion of the whole. And that metaphor might aptly be transferred to the case of justification in general.[42]

Others however hold that in justificatory regresses transitivity fails:

> [regressive transitivity] will often fail — for example in the much-discussed regress of reasons. For ...A_2 can afford a good reason for A_1's acceptance, and A_1 for q's, without A_2 being a good reason to accept q.[43]

To complicate the matter still further, contemporary epistemologists discussing infinitism generated their own paradigm cases. One involves the analogy with basketball players throwing around the ball:

---

[41] Ibid., 520. Brown hypothesizes that the concept of responsibility has something to do with it. Calling in mind the etymology of 'cause' (which goes back to the Greek 'aitia', a term that occurs mainly in legal contexts), Brown argues that it is precisely the connotation of 'cause' as something that is *responsible* for its effect that is crucial here: an essentially ordered series needs a first member because it needs a member that is responsible for the entire series.

[42] Hankinson 1995, 189.

[43] Rescher 2010, 83, footnote 1. We have changed the symbols so as to make them match ours.

1.3 Vicious Versus Innocuous Regress

> Consider the analogy of basketball players ... passing the ball to another. ...the question is this: how did [the ball] get there in the first place? [44]

Here epistemic justification which goes from one proposition to another is compared to a ball that is passed from one player to another. Is this a helpful picture? Not so sure: the picture suggests that justification is something that is lost once it is handed over to the neighbouring proposition, and this is not something that we associate with justification. We do not believe that, if A_i justifies A_j, the former thereby loses the property of being justified — quite the opposite. In this respect justification seems more like dharma transmission or like an infectious disease: holy man A_i can impart dharma to person A_j without losing his holiness, just as the sick person A_i can pass on his infection to person A_j without thereby being cured. [45]

In logic and mathematics, a necessary condition for establishing whether a series continues indefinitely is to know the domain and the relation in question.[46] Take the formula $\forall x \exists y Rxy$. Whether this formula is true or false depends on the domain over which the variables x and y range and on the nature of the relation R. That is, we need to know what the objects in the series are and also what the relation between those objects is. The statement S: 'For all objects x there is an object y such that y is smaller (or less) than x' is true if x and y are integers; S then unproblematically covers an infinitude of objects. But if x and y are natural numbers, then S is false, since there is a smallest natural number. However, if we change S into S': 'For all objects x there is an object y such that y is *greater* than x', then we obtain a truth even with the interpretation of x and y as natural numbers. This illustrates that not only the character of the objects is important, but the nature of the relation between the objects too.

As do the causal cases, these mathematical considerations intimate that, also in the field of epistemic justification, we must at least make clear what the meaning is of the A_n, the objects, and of ⟵, the arrow which symbolizes the relation between the objects. What are reasons in a justificatory chain? And how are they related? Only after having settled these questions could

---

[44] Klein 2011b, 494.

[45] John Turri also noted that 'justification' does not imply that something gets lost (Turri 2014, 222). However, he uses the word 'transmission' for the latter case. In Turri's terminology, if a property gets transmitted from A_j to A_i, this means that A_j loses the property while A_i receives it. Our use of 'transmission' is different, in that it does not imply that A_j no longer has the property.

[46] *Cf.* Beth 1959, Chapter 1, Section 4.

we hope to assess whether a justificatory chain of infinite length is sensible or nonsensical, and we will address these matters in the next chapter.

Open Access This chapter is licensed under the terms of the Creative Commons Attribution 4.0 International License (http://creativecommons.org/licenses/by/4.0/), which permits use, sharing, adaptation, distribution and reproduction in any medium or format, as long as you give appropriate credit to the original author(s) and the source, provide a link to the Creative Commons license and indicate if changes were made.

The images or other third party material in this chapter are included in the chapter's Creative Commons license, unless indicated otherwise in a credit line to the material. If material is not included in the chapter's Creative Commons license and your intended use is not permitted by statutory regulation or exceeds the permitted use, you will need to obtain permission directly from the copyright holder.

Chapter 2
Epistemic Justification

Abstract
What is the nature of the justifier and of the justified, and how are they related? The answers to these questions depend on whether one embraces internalism or externalism. As far as the formal side of the justification relation is concerned, however, the difference between internalism and externalism seems irrelevant. Roughly, there are three proposals for the formal relation. One of them conceives the justification relation as probabilistic support; in fact, however, probabilistic support is only a necessary and not a sufficient condition for justification.

2.1 Making a Concept Clear

In philosophy concepts typically resist definition. Truth, justice, beauty, freedom, goodness: each of these notions is as fundamental as it is enigmatic. It has been argued that the perennial attempts to define these terms are part and parcel of the philosophical game, setting philosophy apart from science. Thus Kant maintained that the way in which philosophers define their concepts differs essentially from the way in which definitions are given outside of philosophy. Defining a term in mathematics or the sciences, as he writes in his *Logic*, is "to make a clear concept" whereas defining a philosophical concept is "to make a concept clear".[1] Giving a definition of the mathematical term 'trapezium', for example, amounts to combining previously existing and supposedly unambiguous notions like 'parallel', 'angle', and 'side' into

---
[1] *Einen deutlichen Begriff machen* versus *einen Begriff deutlich machen* — Kant, Logik, Einleitung, VIII C, see Jäsche 1869/1800, 70.

the newly constructed concept 'trapezium'. In defining a philosophical term, however, we do not construct a new term out of already existing elements, but rather try to reconstruct and clarify what is already given to us in a confused and ill-determined way. While in philosophy we have a vague understanding of the *definiendum* and strive to come up with a *definiens* that agrees with what we have in mind, in the sciences we fabricate a *definiendum* on the basis of a clear and existing *definiens*.

Although this Kantian view of philosophy dates back to Plato and was still upheld in the twentieth century by such major figures as Rudolf Carnap, it is all but uncontroversial. Especially in the late twentieth century pragmatists and naturalists blamed it for its alleged sterility, and for its failure to appreciate the continuity between science and philosophy. Richard Rorty expresses the point unreservedly:

> Pragmatists think that the story of attempts ...to define the word 'true' or 'good' supports their suspicion that there is no interesting work to be done in this area. It might, of course, have turned out otherwise. People have, oddly enough, found something interesting to say about the essence of Force and the definition of 'number'. They might have found something interesting to say about the essence of Truth. But in fact they haven't. ... [P]ragmatists see the Platonic tradition as having outlived its usefulness.[2]

Our aim in this chapter is to say something interesting about the concept of epistemic justification. The subject has been in a predicament ever since Plato in his *Theaetetus* set out to answer the question: 'What is the distinction between knowledge and true belief?' Nowadays the debate about epistemic justification has become a multi-faceted affair, consisting of many sub- and sub-sub-discussions. More than once the participants in the debate have crossed the borders of epistemology in order to continue their arguments in the fields of ethics, metaphysics, or philosophy of mind, thus creating a complicated and colourful network of various positions with myriad connections and interdependencies.

Some are pessimistic that progress can be achieved here. Roderick Chisholm discouragingly comments on Plato's undertaking in the *Theaetetus*: "It is doubtful that he succeeded and it is certain that we cannot do any better."[3]. William Alston, as we have seen, is even more explicit. He firmly recommends the abandonment of the project of trying to understand justification altogether, and makes a plea for an epistemology that studies various 'epis-

---

[2] Rorty 1982, xiv.
[3] Chisholm 1966, 5.

2.1 Making a Concept Clear

temic desiderata'. Like Rorty, he is clearly annoyed with a project that has demanded so much from so many and has delivered so little.[4]

Our concern with epistemic justification in this book is in fact secondary. When we aim to say something interesting about the issue, this is not because we aspire to define epistemic justification as such. Rather, we want to find out whether, and if so to what extent, it makes sense to speak of infinite justificatory chains. The major proponent of infinite chains in epistemology, Peter Klein, has argued that the latter objective can be achieved without the former. As he sees it, we need not unduly exert ourselves to understand justification, for the meaning of epistemic justification is irrelevant to a discussion about the possibility of infinite justificatory chains. We can merrily discuss the pros and cons of infinitism without worrying about what exactly justification means, since infinitism is consistent with the many different accounts of the expression 'A_j justifies A_i'. Klein lists five of those accounts, adding that the list is not exhaustive:

1. if A_j is probable, then A_i is probable and if A_j is not probable, then A_i is not probable; or
2. in the long run, A_j would be accepted as a reason for A_i by the appropriate epistemic community; or
3. A_j would be offered as a reason for A_i by an epistemically virtuous individual; or
4. believing that A_i on the basis of A_j is in accord with one's most basic commitments; or
5. if A_j were true, A_i would be true, and if A_j were not true, A_i would not be true.[5]

---

[4] Alston's exasperation is rooted in his belief that "[t]here isn't any unique, epistemically crucial property of beliefs picked out by 'justified'" (Alston 2005a, 22). *Cf.* "...it is a mistake to suppose that there is a unique something-or-other called 'epistemic justification' concerning which the disputants are disputing" (Alston 1993, 534). See also Alston 1989, where similar ideas are defended. Likewise, Richard Swinburne has denied that there exists one pre-theoretic concept 'epistemic justification', which can subsequently be made clear in the way that Plato and Kant proposed (Swinburne 2001). But where Alston advises us to withdraw from justification research and to stop talking about justification altogether, Swinburne encourages us to let a thousand justificatory flowers bloom. The two positions may seem to be poles apart, but in a sense there is much overlap. In the end, Swinburne's pluralistic view of epistemic justification and Alston's plea for a plurality of epistemic desiderata might perhaps differ only terminologically.

[5] Klein 2007a, 12. Klein has p and q instead of A_j and A_i.

He concludes that "infinitism can opt for whatever turns out to be the best account since each of them is compatible with what infinitism is committed to."[6] As we have indicated in the previous chapter, we think this is too quick. Whether an infinite regress of justification makes sense may very well depend on the meaning of justification. We have seen that under a particular interpretation of 'x causes y' or 'x is smaller than y' regresses may be harmless, whereas under an alternative interpretation they do not make sense. Something similar might well apply to the case of epistemic regresses, so it is incumbent upon us to consider what may be meant by 'A_j justifies A_i'.

In doing so, we are not trying to give a definition of justification. Nor are we making any claim about its relation to knowledge. Traditionally, justification has been seen as a necessary ingredient of knowledge, as that which has to be added to true belief. Recently, different views have been put forward, such as that justification is a derivative of the primitive concept of knowledge, or is possible knowledge, or potential knowledge, or appearance of knowledge, or that it implies truth, or that justification and knowledge simply coincide.[7] We will make no such claims. Everything we say about justification is meant as a contribution to the debate about the possibility of epistemic regresses, not to the debate about how to define knowledge or justification. Of course, the two issues are connected, but it would be a mistake to treat them as being on a par.[8] Our aim, as said, is to find out to what extent infinite epistemic chains are possible. It will turn out that for this purpose it is enough to adopt a very modest and uncontroversial claim about justification; there is no need to define justification or to say how exactly it relates to knowledge.

---

[6] Ibid.

[7] Williamson 2000; Bird 2007; Jenkins Ichikawa 2014; Reynolds 2013; Littlejohn 2012; Sutton 2007. We will shun the term 'warrant' when we speak of justification, since some reserve this term for 'that which added to true belief yields knowledge'. See Plantinga 1993.

[8] That a theory of justification is different from a theory of knowledge has been argued in Booth 2011 and Foley 2012. Alvin Goldman also acknowledges that an interest in justification can have several different motivations, only one of which is an interest in knowledge as such (Goldman 1986, 4). Martin Smith, however, defends what he calls "the normative coincidence constraint", according to which aiming at justification and aiming at knowledge coincide. We will say more about Smith's views in 2.5.

2.2 Two Questions

The first thing to note when considering the concept of epistemic justification is that it is a relational notion, corresponding to a two-place predicate. When we say that something is justified, we mean that it is justified *by something else*. This 'something else' can be of the same ontological category as the thing justified, as for example when a belief justifies another belief. But it can also belong to another category, as when we say that a belief is justified by an experience or by an event. It might also happen that something justifies itself, and then the justifier and the thing justified coincide. My belief that I am having a belief, for example, falls into the latter category.

Once the relational character of justification has been acknowledged, we can appreciate that the question 'What is justification?' in fact consists of two questions. If we want to know what is meant by the expression 'A_j justifies A_i', we will have to answer both $Q1$ and $Q2$:

$Q1$: What is the character of the relata, A_j and A_i?

$Q2$: What is the character of the relation between A_j and A_i?

The difference is clear.[9] $Q1$ is about the ontology of reasons. It is a question about the stuff that the objects A_i and A_j are made of. Are they abstract entities like propositions? Psychological entities like beliefs? Or are they events, or facts, or material objects? Question $Q2$ is about their connection. In the previous chapter we symbolized this connection by a single arrow, but what does this symbol mean? Is it the arrow of entailment, as has been argued by for example James Cornman and John Post?[10] Does it represent 'probabilification', as William Alston and Matthias Steup have called it?[11] Is the

---

[9] Others have also stressed this difference. Andrew Cling, for example, writes that "a theory of ... reasons ... must do two things. First, it must give an account of the *relationship* that must obtain between ... reasons and their specific targets. ... Second, [it] must specify the characteristics that a mental state must have if it is to be a reason for *any* target." (Cling 2014, 62). Similarly, Ram Neta distinguishes between "reasons in the light of which a claim is justified" and "the relation ... between those reasons" (Neta 2014, 160). The very distinction is also central to Richard Fumerton's 'Principle of Inferential Reasoning', that we will discuss in Section 8.5.

[10] Cornman 1977; Post 1980.

[11] Alston 1993, 528; Steup 2005, Section 2.1. Regarding the question about the relation between A_i and A_j, Michael Williams takes a very different view. As he sees it, "there is no relation to account for" and he comments further: "There may well be relations of entailment ... or conditional probability .... But no such relation

justification relation primarily a logical relation, as is stated by Richard Feldman and Earl Conee?[12] Or should we follow David Armstrong and Alvin Goldman and hold that it is ultimately causal?[13]

Which answer one gives to $Q1$ and $Q2$ depends largely on whether one takes an internalist or an externalistic view of epistemic justification. Our intuitive understanding of epistemic justification, which Kant would have called "confused and undetermined", revolves around two aspects that philosophers of all times have struggled to amalgamate.[14] On the one hand, justification has to do with the way the world *is*: it would be inappropriate to call our beliefs justified without requiring that they represent, at least remotely, how things actually are. On the other hand, justification applies to the way the world *appears* to us: it would be awkward to call my beliefs *un*justified if I have reasoned impeccably towards a conclusion which, through some freakish turn of fate, happens to be false. The fact that externalists tend to stress the former, world-centred aspect of justification, while internalists emphasize the latter, agent-centred aspect, is reflected in their answers to $Q1$ and $Q2$.

What do internalists and externalists say about the ontological status of the relata in 'A_j justifies A_i'? Concerning A_i, the thing justified, there seems to be not much disagreement. In the case at hand, both factions assert that A_i is a proposition, or a belief in a proposition.[15] But what about the ontological status of the justifier, A_j? Here the answer depends on which of the many different versions of internalism or externalism we are talking about. It also depends on whether A_j is regarded as something that is itself inferred or as

---

suffices to make a proposition a reason for another" (Williams 2014, 237). As will become clear later in this chapter, we agree that a relation of entailment or of conditional probability is not sufficient for saying that one proposition is a reason for another. However, if Williams is implying that such a relation is not necessary either, then we part company. In general, Williams' approach to the epistemic regress problem is inspired by the later Wittgenstein and by ordinary language philosophers like Austin, and as such tends to eschew a more formal or theoretical approach, like the one that we pursue in this book.

[12] Feldman and Conee 1985; Conee and Feldman 2004.

[13] Armstrong 1973; Goldman 1967.

[14] *Verworren und unbestimmt* — see Kant's treatise 'Enquiry concerning the clarity of the principles of natural theology and ethics' (*Untersuchung über die Deutlichkeit der Grundsätze der natürlichen Theologie und der Moral*) of 1764. *Cf.* Vahid 2011. We recall agreeable conversations with Hans Mooij and Simone Mooij-Valk about translating Kant and — *vis-à-vis* the motto of this book — Thomas Mann.

[15] But only in the case at hand, for A_i can also be another cognitive state than a belief, or even a non-cognitive state.

2.2 Two Questions

something that is non-inferential. If A_j is itself inferred, all internalists and some externalists will see A_j too as a belief or a proposition. If A_j is not inferred, some internalists will see it as a belief or a proposition (albeit of a special, basic kind), whereas other internalists maintain that in this case A_j is a fact or an experience. Externalists will regard A_j in this case as a fact, an object, or an event, but they differ in their opinions about what kind of fact, object, or event A_j exactly is. Some say that it is a fact outside us; for example, if an airplane is flying by, and my perceptual and cognitive wiring is as it should be, then this fact is a reason for me to believe that an airplane is flying by. Other externalists hold that it is a fact inside us, for example the activation of my retina or my eardrum, causing neural events and brain states culminating in my belief that an airplane is flying by.

As to $Q2$, internalists tend to be evidentialists: they see the relation between the relata in 'A_j justifies A_i' as being *logical* or *conceptual* in character. For them, A_j is a good epistemic reason for A_i if A_j is *adequate evidence* for A_i. As Earl Conee and Richard Feldman phrase it:

> The evidentialism we defend ... holds that the epistemic justification of a belief is a function of evidence.[16]

> According to evidentialism, a person is justified in believing a proposition when the person's evidence better supports believing that proposition than it supports disbelieving it or suspending judgement about it. ... when a belief is based on justifying evidence, then ... the belief is well-founded.[17]

Externalists, on the other hand, are mostly reliabilists: in their view, A_j is a good epistemic reason for A_i if and only if A_i has been *reliably formed* on the basis of A_j, where a belief-forming method is reliable if it results in acquiring true beliefs and avoiding erroneous ones. Reliabilists see the reliability relation as being *nomological* or even *causal* in nature.[18] They criticize evidentialists for neglecting the difference between logic and epistemology, stressing that, while logic deals with inferences and the validity of argument forms, epistemology has to do with the practice of forming actual beliefs. As one of the pioneering reliabilists has it:

> ... although epistemology is interested in inference, it is not (primarily) interested in inferences construed as argument forms. Rather, it is interested in

---

[16] Conee and Feldman 2004, 2.

[17] Ibid., 3.

[18] One of the first papers that stresses the difference between logical and causal relations is Davidson 1963, although it does not contain the terms 'evidentialism' or 'reliabilism' and is about rational action rather than about justified belief.

inferences as processes of belief formation or belief revision, as sequences of psychological states. So psychological processes are certainly a point of concern, even in the matter of inference. Furthermore, additional psychological processes are of equal epistemic significance: processes of perception, memory, problem solving, and the like.

Why is epistemology interested in these processes? One reason is its interest in epistemic justification. The notion of justification is directed, principally, at beliefs. But evaluations of beliefs ... derive from evaluations of belief-forming processes. Which processes are suitable cannot be certified by logic alone. Ultimately, justificational status depends (at least in part) on properties of our basic equipment. Hence, epistemology needs to examine this equipment, to see whether it satisfies standards of justifiedness.[19]

These standards of justifiedness are given by the "right system of justificational rules" or J-rules.[20] No J-rule can be generated by logic alone, the main reason being that J-rules govern the transitions to states of belief, and that logic is not about such states:

> ... logic formulates rules of inference, which appear in both axiomatic systems and natural deduction systems. But these rules are not *belief*-formation rules. They are simply rules for writing down formulas. Furthermore, formal logic does not really endorse any inference rules it surveys. It just tells us semantic or proof-theoretic properties of such rules. This is undoubtedly *relevant* to belief-forming principles ... But it does not in itself tell us whether, or how, such rules may be used in belief formation.[21]

In the end, Goldman opts for what he calls "the absolute, resource-independent criterion of justifiedness":

> A J-rule system R is right if and only if R permits certain (basic) psychological processes, and the instantiation of these processes would result in a truth ratio of beliefs that meets some specified high threshold (greater than .50).[22]

Evidentialism and reliabilism are usually described as opposing positions, but recently arguments have been put forward for a rapprochement between the two, including some arguments by Goldman himself.[23] We similarly ad-

---

[19] Goldman 1986, 4.
[20] Ibid., 59.
[21] Ibid., 82.
[22] Ibid., 106. Goldman adds that the rightness of the rule system R should be in the set of "normal worlds", i.e. "worlds consistent with our *general* beliefs about the actual world". (Ibid., 107.)
[23] Goldman 2011; Comesana 2010. Alston 2005a, Chapter 6, defends the thesis that evidentialism and reliabilism are virtually identical.

vocate a reconciliation, but our argument is different from the existing ones in that it relies on the formal side of the justification relation.

Neither evidentialists nor reliabilists have been very explicit about this formal side. Conee and Feldman, when referring to the relation, speak about 'fittingness': a belief in a proposition is justified for a particular person if and only if that belief fits the person's evidence.[24] They explicitly refrain from describing the fitting relation in formal detail, presumably because they want to keep their analysis as general a possible. Reliabilists, of course, are not particularly interested in the formal side of the justification relation either, since they consider it inessential to the actual process of acquiring a justified belief. In contrast to both groups, we deem it fruitful to investigate the formal structure of the justification relation. As we will see, this will enable us to reconstruct the evidentialist and the reliabilist view as two interpretations of one and the same formal framework.

2.3 Entailment

When it comes to the formal side of the justification relation, we can perceive in the literature three major proposals. According to the first, 'A_j justifies A_i' should be read as

$\qquad$ 'A_j implies A_i' or 'A_j entails A_i'.

According to the second, we should interpret it as

$\qquad$ 'A_j probabilifies A_i' or 'A_j makes A_i probable'.

The third proposal is based on work by Fred Dretske and Robert Nozick, and it is sometimes referred to as truth-tracking.[25] Roughly, it states that a person is justified in believing proposition A_i if this person tracks the truth of A_i, which in this case means: bases his belief in A_i on A_j. On a formal level, the truth-tracking approach makes use of subjunctive conditionals of the form

(a) 'if A_j were the case, then A_i would be the case'
(b) 'if A_j were not the case, then A_i would not be the case'.

---

[24] Conee and Feldman 2004, Chapter 4.
[25] Dretske 1970, 1971; Nozick 1981.

Nozick argues that the subjunctive conditional (a) leads to the probability $P(A_i|A_j) = 1$, and that conditional (b) leads to $P(\neg A_i|\neg A_j) = 1$, or equivalently $P(A_i|\neg A_j) = 0$. This corresponds to what he calls 'strong truth tracking'. As he rightly comments, "the evidence we have for hypotheses is not usually strong evidence; too often although the evidence would hold if [the hypothesis] were true, it might also hold if [the hypothesis] were false."[26] He is then led to a probabilistic approach that is the same as the second proposal. However, rather than framing the subjunctive conditionals in probability statements, it may be more natural to couch them in the language of possible world semantics, invoking David Lewis's method of nearby possible worlds.[27]

We will say something about the first proposal in the present section. The second one will be discussed in Section 2.4, where we argue that probabilistic support is a necessary but not a sufficient condition for justification. We will say more about the third proposal in Section 2.5, where we consider an argument by Martin Smith that can be seen as an objection to our argument in 2.4.

The idea that justification has something to do with implication or entailment appears to be widely accepted. Aristotle assumes it in his writings on epistemic regresses, and many epistemologists in the twentieth century who write about justification seem to have had implication in mind. 'Seem', for the idea often remains implicit. This goes for the literature on epistemic justification in general, but also for the more specific papers on the regress problem in epistemology. For example, Tim Oakley develops an argument according to which no beliefs can be justified since that would require an infinite regress. Before presenting his argument, he writes:

> I offer no analysis of the term 'justified', since this is not required for my argument, and take the notion to be a commonsense one, regularly though unreflectively used by us all.[28]

Oakley's paper makes it however very clear that at least part of his argument only works when justification is taken as implication or as deductive inference. Thus Scott Aikin rightly notes that in Oakley's argument "deductive inference rules play the role of inferential justification".[29] And Oakley is no exception here. Among authors who defend the sceptical position that no belief can be justified because that would demand an infinite epistemic chain,

---

[26] Nozick 1981, 250.
[27] Williamson 2000; Pritchard 2005, 2007, 2008; Sosa 1999a, 1999b.
[28] Oakley 1976, 221.
[29] Aikin 2011, 59; Oakley 1976, Sections 4.3 and 5.3.

2.3 Entailment

many tacitly make the assumption that 'A_j justifies A_i' means 'A_j implies A_i'.

Occasionally, however, authors *are* explicit about their use of the justification relation as implication or entailment. John Post is a case in point.[30] Post first takes justification to be inferential justification and then notes:

> If anything counts as an inferential justification relation, logical implication does ...
> provided it satisfies appropriate relevance and noncircularity arguments.[31]

More particularly, Post sees justification as "proper entailment":

> Let us say a statement A_j *properly entails* a statement A_i iff A_j semantically entails A_j, where the entailment is relevant and non-circular on any appropriate account. Thus if anything counts as an inferential justification relation, proper entailment does, in the sense that where A_j and A_i are statements rather than sets of statements: 'If A_j properly entails A_i, then A_i is justified for [a person] P if A_j is — provided P knows that the proper entailment holds and would believe A_i in the light of it if he believed A_j.[32]

There exist many cases of proper entailment *à la* Post. The example that he himself presents is based on *modus ponens*. If A_j is

$$p \wedge (p \rightarrow q),$$

and A_i is q, then A_j properly entails A_i.

To regard justification as implication or entailment has the advantage (if it is one) that justification is transitive and truth-conducive. However, it has been rightly criticized as a view that puts very strong requirements on the notion of justification, and may typically lead to scepticism if rigorously implemented. In 1978 Richard Foley had already made a plea for allowing "non-paradigmatically justified beliefs", i.e. beliefs of which the justification is not subject to such strong requirements as those that follow from straightforward implication.[33] Foley leaves open what exactly he means by

---

[30] Post 1980. In this paper Post describes a particular objection to infinite regresses that we will discuss in Section 6.3. Post's argument can be seen as an improved version of objections that have been raised by John Pollock and James Cornman (Pollock 1974; Cornman 1977).

[31] Post 1980, 33.

[32] Ibid. We have replaced Post's X and Y by our A_j and A_i. Post talks about "statements" where we use 'propositions' or 'beliefs'. In this chapter we will not distinguish between the latter two terms.

[33] Foley 1978, 316.

non-paradigmatic justification, refraining from giving a general account of the phenomenon, and even doubting whether such an account can be given at all.

The second proposal for a formal rendering of 'A_j justifies A_i', to be discussed below, is more important and more realistic; in fact, it goes some way towards specifying the non-paradigmatic account of justification that Foley has been looking for.

2.4 Probabilistic Support

A distinction is often made between deontological and non-deontological justification. In the deontological understanding, as Matthias Steup phrases it, a person S "is justified in believing that r if and only if S believes that r while it is not the case that S is obliged to refrain from believing that r."[34] Steup notes that the deontological concept "is common to the way philosophers such as Descartes, Locke, Moore and Chisholm have thought about justification", but that today it is deemed "unsuitable for the purposes of epistemology". What *is* deemed suitable today is the non-deontological view, which conceives justification as "probabilification":

> What does it mean for a belief to be justified in the non-deontological sense? Recall that the role assigned to justification is that of ensuring that a true belief isn't true merely by accident. Let us say that this is accomplished when a true belief instantiates the property of *proper probabilification*. We may, then, define non-deontological justification as follows:
>
>> [Person] S is justified in believing r if and only if S believes that r on a basis that properly probabilifies S's belief that r. [35]

Instead of 'probabilification', epistemologists also use the term 'to make probable' for justification. Says for example Richard Fumerton:

> Can we find a way of characterizing epistemic justification that is relatively neutral with respect to opposing analyses of the concept? As a first stab we

---

[34] Steup 2005. Steup has p for our r.

[35] Ibid. The importance of probabilification has also been stressed earlier by William Alston, albeit not as a way of understanding justification, but as one of the epistemic desiderata that deserve thorough study: "The reason or its content must be so related to the target belief and its content that, given the truth of the former, the latter is thereby likely to be true. The reason must sufficiently 'probabilify' the target belief." (Alston 1993, 528).

2.4 Probabilistic Support

might suggest that whatever else epistemic justification for believing some proposition is, it must make *probable* the truth of the proposition believed. The patient with prudential reasons for believing in a recovery was more likely to get that recovery as a result of her beliefs, but the *prudential reasons* possessed did not increase the probability of the proposition believed — it was the belief for which the person had prudential reasons that resulted in the increased probability. Epistemic reasons make likely the truth of what is supported by those reasons ....[36]

Here we shall work under the assumption that 'probabilification' or 'making probable' is essential for the concept of justification. To say that A_j makes A_i probable at least means that A_j raises the probability of A_i if A_j is true, as compared with the value it would have had if A_j had been false. So

$$P(A_i|A_j) > P(A_i|\neg A_j), \tag{2.1}$$

in words: A_i is more probable if A_j is the case than if A_j is not the case.[37] We say that A_j makes A_i more probable if and only if (2.1) is fulfilled. Here we assume that $P(A_j)$ lies strictly between zero and one, but in later chapters we will see how to drop this assumption.

We will call (2.1) the *condition of probabilistic support*. It is in fact equivalent to the classificatory version of what Rudolf Carnap in the preface to the second edition of his *Logical Foundations of Probability* calls "increase in firmness".[38] While Carnap's concept of "firmness" is concerned with *how probable* A_i is on the basis of A_j, his notion of "increase in firmness" relates to the question as to whether and by how much the probability of A_i is *increased* by the evidence A_j. Carnap specifies, both for firmness and for increase in firmness, three versions: a classificatory, a comparative and a quantitative variant. In the classificatory variant of increase in firmness, A_i is *made firmer* by A_j. Or in Carnap's formulation (where we have replaced his c by our P):

$$P(A_i|A_j) > P(A_i|t). \tag{2.2}$$

Here t is the tautology, so (2.2) is the same as

$$P(A_i|A_j) > P(A_i), \tag{2.3}$$

---

[36] Fumerton 2002, 205.

[37] That Fumerton has in mind 'making more probable' or 'increasing the probability' when he writes about 'making probable' is indicated by his use of the expression "the increased probability".

[38] Carnap 1962, xv-xvi.

which is equivalent to our condition of probabilistic support, (2.1), for while $P(A_i|\neg A_j)$ and $P(A_i)$ will in general not be equal to one another, the two inequalities (2.1) and (2.3) imply one another.[39] Three observations should be made about this condition (2.1). First, the condition is quite weak. It only says that A_i is made more probable by A_j than by $\neg A_j$. It does not say that A_i is made more probable by A_j than by another proposition A_k, nor does it claim anything about propositions different from A_i that are made even more probable by A_j. Our condition is silent about the amount of probabilistic support that A_i receives from A_j as compared to the amount of probabilistic support that A_i would have received from another proposition A_k. So the condition does not imply that the former amount is greater than the latter, nor does it imply that the amount of probabilistic support given to A_i should exceed a particular threshold.[40]

Second, the condition of probabilistic support is not a measure. As Branden Fitelson has emphasized, there are many different measures of probabilistic support or confirmation, and they are often ordinally inequivalent to one another.[41] This might be problematic in many contexts, but it is not an issue for us. For the various measures of probabilistic support all agree in stating that A_j probabilistically supports A_i if and only if $P(A_i|A_j) > P(A_i|\neg A_j)$, and this is all we need here.

Third, the condition does not need a threshold. It could reasonably be objected that the phrase 'making probable' involves more than 'making more probable'. If A_j makes A_i probable, then surely the effect of A_j on A_i must be not merely to raise the probability of A_i, but also to raise it above one half (or perhaps above some agreed-upon threshold greater than a half). In Section 6.5 we shall say some more about thresholds, but the chief thing to realize is that a threshold condition is not needed for our purpose: a threshold is not required for finding out to what extent infinite justificatory chains make

---

[39] From the definition of conditional probability it follows that

$$P(A_i|A_j) - P(A_i) = P(\neg A_j)[P(A_i|A_j) - P(A_i|\neg A_j)].$$

The right-hand side of this equation is greater than zero if (2.1) is true, so the left-hand side must be greater than zero too, and this implies (2.3). By similar reasoning, it is clear that, if (2.3) is true, then (2.1) is true. Recall that here $P(A_j)$ is neither 0 nor 1.

[40] We will assume, however, that $i \neq j$. We are after all interested in probabilistic support as a condition of epistemic justification, and in accordance with Peter Klein's Principle of Avoiding Circularity (PAC), A_i may not be in its own evidential ancestry.

[41] Fitelson 1999.

2.4 Probabilistic Support

sense. The only thing we need for this is that (2.1) is to be regarded as a minimal condition implicit in the notion of 'making probable'.

We consider condition (2.1) to be an essential ingredient of the relation of epistemic justification, insofar as it involves probabilification or 'making probable'. It is important to keep in mind that the condition itself is completely formal. It does not imply anything about the ontological character of the relata nor does it make any assumption about how the probability relation should be interpreted. If A_i is a belief or proposition that is justified by A_j, then the justifier A_j can be anything: a belief, a proposition, a fact, an event, a perception, a memory or a neural state — this is completely irrelevant. And P can also be anything: subjective or objective or logical probability, that does not matter. What *does* matter is that (2.1) is governed by the formal probability calculus, i.e. the axioms of Kolmogorov and the theorems that follow from them.[42]

Precisely because the calculus is formal and thus uninterpreted, condition (2.1) is neutral with respect to internalism and externalism, and also with respect to evidentialism and reliabilism. The condition can be combined with internalistic and externalistic views concerning the ontology of reasons, as well as with evidentialist and reliabilist understandings of the justification relation. After all, internalists, externalists, evidentialists and reliabilists do not differ about the probablity calculus, nor is there anything in their positions that goes against formalizing 'A_i is made probable by A_j' in terms of (2.1). The only differences between them are about interpretations: whereas internalists construe A_i and A_j internalistically, externalists construe them externalistically, and whereas evidentialists interpret the probability relation in logical terms, reliabilists interpret it in nomological terms, typically in terms of probabilistic causality. But these are just differences in interpretations, and they do not touch the underlying formal level.

---

[42] If $P(A_i|A_j) < P(A_i|\neg A_j)$, then it is $\neg A_j$ rather than A_j that supports A_i probabilistically. The point that the negation of one event could be the cause of another was already made by Hans Reichenbach when he introduced his concept of the common cause. Reichenbach notices that in this case we must revise our opinion on the working cause, or as he puts it succinctly "in this case A_j and $\neg A_j$ have merely changed places" (Reichenbach 1956, 160 — we have replaced his C by our A_j). Thus an arbitrary chain of probabilistic relations can be recast in a form in which probabilistic support (or neutrality) holds all along the chain. This does not mean that *any* proposition can be justified by *any* probabilistic chain, for the condition of probabilistic support is only a necessary, not a sufficient condition of justification. In Section 6.5 we will look at some further desiderata for an adequate description of what justification entails.

It might be true that more people are inclined to interpret (2.1) along the lines of internalism and evidentialism.[43] Under that interpretation, A_i and A_j are both beliefs or propositions, and P is construed as subjective or epistemic or logical probability. The point we want to make here is that (2.1) can just as easily be understood in accordance with externalism and relabilism. Under this interpretation, A_i and A_j can be beliefs, perceptual appearances, memories, and so on. For example, if A_i is my belief in the proposition that a cow is grazing in front of me, and A_j is my seeing a cow grazing in front of me, then (2.1) states that it is more likely that my belief in a grazing cow is true, given that I have this perception, than when I do not have this perception. Here P is an objective probability, depending on the frequency of events, where the events are 'seeing a grazing cow' and 'believing that there is a grazing cow'. Whether or not (2.1) holds here is determined by empirical research or, more generally, by past performance. Is it the case that my seeing a cow grazing is more often followed by a belief in a cow grazing than that my perception of a horse jumping is followed by a belief that a cow is grazing? The answer is presumably in the affirmative, so (2.1) is satisfied.

The thing justified need not be a belief of which the probability is determined on the basis of perceptual appearances. It can also be the other way around. In cases of wishful thinking or of harbouring strong suspicions, my beliefs or my desires can cause in me certain perceptual appearances. Here the causal course runs in the opposite direction. Again, we determine empirically whether or not a causal process is in fact taking place, and thus whether or not (2.1) is satisfied: some people are more prone to wishful thinking or to being suspicious than others.

It might happen that a causal process gives rise to a false belief. Optical illusions are a classic example. When I am walking in the desert, the refraction of light from the sky by heated air can cause me to believe that there is a sheet of water in front of me. This causal process is probabilistic (at some times I am more vulnerable to this optical illusion than at others), but it is assumed that (2.1) is fulfilled (I mostly fall prey to the illusion when walking in the desert). Goldman will presumably say that this is not a reliable belief-forming process, and that my belief that there is a sheet of water in front of me is not justified. As we will explain in the next section, however, we are not proposing to *define* justification as the condition of probabilistic support

---

[43] Thus René van Woudenberg and Ronald Meester have argued that "the traditional epistemic regress problem" is by definition cast in internalistic and doxastic terms, and they seem to hold that it should also be considered in those terms (Van Woudenberg and Meester 2014). For more on internalism and the regress problem, see Simson 1986 and Jacobson 1992.

2.4 Probabilistic Support

(2.1), nor are we saying that, if the condition is satisfied, then there is justification. Our claim is only the moderate one that the condition of probabilistic support is a necessary ingredient of epistemic justification. This claim in no way conflicts with a reliabilist stance *à la* Goldman.[44]

In sum, our condition of probabilistic support is neutral with respect to many debates about justification. One may understand justification internalistically or externalistically, in accordance with evidentialism or with reliabilism — all these views can be combined with the condition of probabilistic support as a mechanism underlying justification. And there are more views that we can accommodate. For example, the condition of probabilistic support is consistent with either side in the debate about the difference between diachronic and synchronic justification, to which especially Swinburne has drawn attention; the only thing we have to do to account for this difference is to add a time index to, or remove it from (2.1).[45] Furthermore there is no reason to restrict the causal processes modelled by (2.1) to individual people; we might well regard them as taking place in a community. Alvin Goldman writes:

> The task of social epistemology ... is to evaluate truth-conducive or truth-inhibiting properties of such relationships, patterns, and structures. What kinds of channels, and controls over channels, comprise the best means to 'verific' ends? To what degree should control be consensual, and to what degree a function of (ascribed) expertise, or 'authority'? To what extent should diversity of messages be cultivated?[46]

---

[44] While the application of (2.1) to internalism is straightforward, since the probability space is homogeneous (containing only propositions or beliefs), the application to externalism is a bit more complicated. Within externalism many things can be reasons, so the probability space is rather diverse, containing not only beliefs and propositions, but also perceptions, memories, facts, and so on. This difficulty can however be handled as follows. First we define different spaces: a space of beliefs, a space of perceptions, a space of memories, and so on. Then we define a space which is the Cartesian product of all those spaces. And finally we decide which relation of probabilistic causality we want to focus on. Do we want to focus on perceptions causing beliefs? Or memories causing beliefs? Or beliefs causing desires? Desires causing beliefs? Deciding on the answers to these questions is necessary in order to keep a grip on the heterogenous probability space, but it is just a slight technical complication, and it is not important for the general philosophical point that (2.1) is neutral with respect to both internalism and externalism.

[45] Swinburne 2001, Chapters 2, 7, and 8.

[46] Goldman 1986, 5-6.

Because the probability calculus is neutral with respect to the nature of the relationships, patterns, and structures that Goldman is talking about, it can accommodate social epistemology alongside individual epistemology.[47]

Some philosophers have doubted that our condition of probabilistic support is in fact neutral. Richard Fumerton first presents the following as a neutral, preliminary characterization of epistemic justification

> epistemic justification ... must make *probable* the truth of the proposition believed,

and then comments:

> Our preliminary characterization of justification as that which makes probable the truth of a proposition may not in the end be all that neutral.[48]

Fumerton gives two reasons why the 'making probable' relation may after all not be neutral. The first is that a "normative feature of epistemic justification ... may call into question the conceptual primacy of probability as a key to distinguishing epistemic reasons from other sorts of reasons".[49] The second is that "if one understands the relation of making probable in terms of a frequency conception of probability, one will inevitably beg the question with respect to certain internalist/externalist debates over the nature of justification".[50] Let us briefly look at each of these two reasons.

The idea behind the first is that epistemic reasons differ from moral or prudential or legal ones, since an epistemic goal is not the same as a goal

---

[47] The condition of probabilistic support is also neutral with respect to several desiderata for the justification relation. For example, Oliver Black required that the relation be irreflexive and transitive (Black 1988; for a "more frugal" formulation of the desiderata, see Black 1996); Romane Clark required transitivity and asymmetry (Clark 1988, 373); and Andrew Cling has argued that having both transitivity and irreflexivity is too strong as a desideratum, proposing an "improved version" of the epistemic regress problem (Cling 2008; for critical replies to Cling, see Kajamies 2009 and Roche 2012; for a further discussion about the transitivity of justification see Post and Turner 2000 versus McGrew and McGrew 2000).

Our claim that probabilistic support is necessary for justification does not preclude justification's being irreflexive or transitive or asymmetrical, even though probabilistic support itself is reflexive, not transitive, and symmetrical. In other words, the claim is not in conflict with the above desiderata for the justification relation, but it does not necessitate them. We will come back to this point at the end of Chapter 6.

[48] Fumerton 2002, 205.
[49] Ibid., 205-206.
[50] Ibid., 206.

2.4 Probabilistic Support

that is moral, prudential, or legal. Whenever one believes a proposition for epistemic reasons, one believes that proposition because it is *probably true*, not because it is useful or moral to believe it. But suggestive as this account of an epistemic reason may be, says Fumerton, "we are in danger of collapsing the distinction between true belief and justified belief".[51] For if a belief is justified if and only if it is probably true, then "our 'goal' oriented account of epistemic justification becomes pathetically circular".[52]

Fumerton's worry can however be dispelled. In order to credit the important role of probabilistic support for justification, we need not say that a belief is probably true if and only if it is justified. It is enough to say that being probably true is a necessary ingredient of being justified. It seems to us that Fumerton is confusing the conceptual *primacy* of probability for justification with its *sufficiency*. He would have been right if 'probabilistic support' had been taken as sufficient for epistemic justification, but as we have said, there is no need to do so. Probabilistic support is merely a necessary, and by no means a sufficient condition for justification.[53]

What about Fumerton's second reason? He is certainly right that the 'making probable' relation will support externalistic and undermine internalistic positions if understood in frequency terms. But our point is that we need not understand the relation in frequency terms, nor need we understand it in nonfrequency terms. If we regard the relation at its formal level, as the condition (2.1), then we are not committed to any interpretation. While Fumerton regards the 'making probable' relation as something that comes with an interpretation, we have construed it as a mere formal structure with uninterpreted symbols.

Apparently Fumerton sees the poly-interpretability of the calculus as a drawback for the idea of modelling justification by means of probability theory. This is the view not only of Fumerton, the internalist, but also of Goldman, the externalist:

> Another admissible theory would let justifiedness arise from the corpus of a cognizer's beliefs plus probabilistic relationships between the target beliefs

---

[51] Ibid., 209.

[52] Ibid.

[53] René van Woudenberg and Ronald Meester appear to think that we deem probabilistic support to be sufficient for justification (Van Woudenberg and Meester 2014). They criticize our condition (2.1) on the grounds that it allows $P(A_i|A_j)$ and $P(A_i|\neg A_j)$ both to be very small, so that $P(A_i)$ is also very small; in that case (2.1) is fulfilled, but it would be ridiculous to say that $P(A_i)$ is justified. The criticism of Van Woudenberg and Meester fails precisely because (2.1) is not a sufficient condition for justification.

and the beliefs in this corpus (or rather, the propositional contents of these beliefs). But here a theorist must tread carefully. The term 'probability' is notoriously ambiguous, and some of its proposed explications implicitly render it a term of epistemic evaluation (tied to what an epistemically rational person would do). For present purposes, 'probability' would have to be restricted to some other meaning, for example, a frequency, or propensity sense.[54]

Goldman is right that the term 'probability' is notoriously ambiguous, but only if we take the term to mean 'calculus plus interpretation', not if it refers to the calculus alone. We are inclined to consider the poly-interpretability of the calculus as an advantage rather than a drawback, since it enables us to work with a well-defined formal framework from which we can derive consequences that hold irrespective of the interpretation. A comparison with Donald Davidson's work might make the point clearer. Davidson argued that an action is only explained (or 'rationalized' as he calls it) if it is both *logically* and *causally* connected to the relevant beliefs and desires.[55] This raises however the question how the two connections can be combined. How to reconcile the position of the so-called causalists with that of the adherents of the Logical Connection Argument, as Frederick Stoutland has aptly called their adversaries?[56] Davidson's own ingenious answer, motivated by his anomalous monism, was that causality is essentially dual. It involves singular causal statements of the form 'token event E causes token event F', which can be true independently of how E and F are described, as well as causal explanations, which centre around causal laws ('events of type $\mathscr{E}$ cause events of type $\mathscr{F}$'), and thus are valid only under certain descriptions.

Resourceful as Davidson's answer may be, a reconciliation between logical and causal connections appears easier once we have taken recourse to probability theory. For probability can model both the logical and the causal relation between reasons and the beliefs or actions that they explain or justify. All we have to do is to replace the logical relations by probabilistic relations, and to substitute probabilistic causality for causality *tout court*. No assumption about a dual character of causality is needed.

As we will show in the chapters to come, the probability calculus has consequences which are very relevant to the possibility and impossibility of infinite epistemic chains. This is not to say that the probability calculus does not face problems, or that its interpretations are unproblematic. It is well known that it has many difficulties that are far from being solved: the problems of

---

[54] Goldman 1986, 24.
[55] Davidson 1963, 1970. *Cf.* our footnote 18.
[56] Stoutland 1970; *Cf.* Peijnenburg 1998.

2.5 Smith's Normic Support

old evidence, spurious relations, irrelevant conjunctions, the prior, the reference class, randomness, and so on. Moreover, apart from those technical quandaries, there is the mundane fact that in actual reasoning the calculus seems to be often violated.[57] These problems are grave indeed, but we do not think they are reasons to reject the calculus as a means of shedding light on the elusive concept of justification. We will say a bit more on this in Section 6.6.

2.5 Smith's Normic Support

We have been arguing that the formal condition of probabilistic support is a necessary ingredient of epistemic justification and, more generally, that Kolmogorovian probability can help us understand what justification is. In this section and in the next one, we will discuss what can be seen as two objections to these views.

The first objection is based on work by Martin Smith. It takes its inspiration from what we have called the third proposal for framing the formal side of the justification relation. According to that proposal, knowledge and justification are best understood on the basis of subjunctive conditionals, which in turn are framed as statements about nearby possible worlds.

Smith has identified a conception of justification which he claims is "taken for granted by a broad range of epistemologists", and which he calls 'the risk minimisation conception of justification':

> ... for any proposition A we can always ask how *likely* it is that A is true, given present evidence. The more likely it is that A is true, the more justification one has for believing it. The less likely it is that A is true, the less justification one has for believing that it is. One has justification simpliciter for believing

---

[57] 'Seems', because sometimes the violation is only apparent. For example, there have been many attempts to explain the famous conjunction fallacy, where people deem the probability of a conjunction ('Linda is a bank teller and a feminist') to be higher than the probability of a conjunct ('Linda is a bank teller'). According to one of these explanations, if we reconstruct the reasoning in terms of confirmation measures rather than of bare probability values, then it is no longer fallacious (Crupi, Fitelson, and Tentori 2007). It is true that we often do not know which mistake exactly people are making, or whether they are making a mistake at all (we might after all have insufficient information about their cognitive make-up). But from this it does not follow that the probability calculus is not a useful instrument to investigate whether mistakes are being made, and, if so, what the nature of these mistakes is.

A (at least at a first approximation) when the likelihood of *A* is sufficiently high and the risk of ¬*A* is correspondingly low. Call this the *risk minimisation* conception of justification.[58]

Describing a proposition as 'likely' means here that the proposition has "an evidential probability that exceeds some threshold *t* that lies close to 1 and may be variable or vague".[59] Smith notes that there is "something very natural" about the entire risk minimization view, but nevertheless concludes that it is "not true at all".[60] According to him it reduces epistemic justification to evidential support, and wrongly so. For it might happen that the evidential support is high, well beyond some threshold of acceptance, while intuitively we would not say that there is justification. Conversely, there might be justification even though the evidential support is relatively low.[61] Smith has illustrated these claims with several appealing examples. Here is one which he borrowed from Dana Nelkin:

> Suppose that I have set up my computer such that, whenever I turn it on, the colour of the background is determined by a random generator. For one value out of one million possible values the background will be red. For the remaining 999 999 values, the background will be blue. One day I turn on my computer and then go into the next room to attend to something else.
>
> In the meantime Bruce, who knows nothing about how my computer's background colour is determined, wanders into the computer room and *sees*

---

[58] Smith 2010, 11. *Cf.* Smith 2016, 2. We have substituted *A* for *P*.

[59] Smith 2016, 29. Note that the concept of evidential probability as Smith uses it here is not the same as the concept of probabilistic support that we talked about in the previous section. Smith says that *A* is likely if and only if its evidential probability, or evidential support, exceeds some threshold: $P(A|E) > t$, where E is the evidence. But we say that $P(A|E)$ satisfies the condition of probabilistic support if and only if $P(A|E) > P(A|\neg E)$.

[60] Ibid., 30.

[61] The difference between epistemic justification and evidential support has been stressed by many others as well. For example, Jarrett Leplin argued that a belief may be highly probable while not justified, and it may be justified even though its probability is very low (Leplin 2009, 101-109). We think that the latter claim is questionable, but even if we grant both claims, Leplin's argument would not affect our view. For Leplin is not talking about probabilistic support in our sense, but about a probability above a certain threshold. The latter also applies to the analysis by Tomoji Shogenji (Shogenji 2012), which we will discuss in Section 6.5. Other defenders of the difference between justification and evidential support are Peter Klein (1999, 312; 2003, 722), Scott Aikin (2011, Chapter 3), and in general proponents of Williamson's 'knowledge first' approach as well as champions of both the safety and sensitivity condition for knowledge.

2.5 Smith's Normic Support

that the computer is displaying a blue background. He comes to believe that it is. Let's suppose, for the time being, that my relevant evidence consists of the proposition that (E_1) it is 99.9999% likely that the computer is displaying a blue background, while Bruce's relevant evidence consists of the proposition that (E_2) the computer visually appears to him to be displaying a blue background.[62]

Let A be the proposition that the computer is displaying a blue background. It is clear that my evidence E_1 does not imply A, since E_1 is compatible with a red background. But neither does E_2 imply A: "After all, Bruce could be hallucinating, or he could be struck by colour blindness, or there could be some coloured light shining on the screen, etc."[63] The point Smith makes is that Bruce's belief in A is a candidate for knowledge, whereas my belief in A is not:

> Bruce's belief would appear to be a very promising candidate for knowledge — indeed, it *will* be knowledge, provided we will fill in the remaining details of the example in the most natural way. My belief, on the other hand, would not constitute knowledge even if it happened to be true. If there were a power failure before I had the chance to look at the computer screen, I might well think to myself 'I guess I'll never *know* what colour the background really was'. But Bruce certainly wouldn't think this.[64]

This means, says Smith, that Bruce is epistemically justified in believing A while I am not. And this is so, even if we assume that A is more likely given my evidence E_1 than given Bruce's evidence E_2:

$$P(A|E_1) > P(A|E_2).$$

The reason why Bruce is justified and I am not, is that the relation between A and E_2 is one of *normic* support, whereas the relation between A and E_1 is only characterized by mere evidential support. Mere evidential support relations only imply that events are likely or unlikely, but normic support relations tell us when events are normal or abnormal:

> Given my evidence E_1, A would frequently be true. Given Bruce's evidence E_2, A would *normally* be true. . . . .[65]

If I were to find out that the background of my computer screen is actually red, I would conclude that this had been merely very unlikely, not that it

---

[62] Smith 2010, 13. *Cf.* Nelkin 2000, 388-389.
[63] Smith 2010, 14.
[64] Ibid.
[65] Ibid., 16.

was abnormal. But if Bruce discovers that the background actually is red, he would conclude that something is not right: he would require an explanation in a way that I would not. In general, E normically supports A if the truth of both E and $\neg A$ is not only unexpected, but a genuine anomaly:

> Say that a body of evidence E *normically supports* a proposition A just in case the circumstance in which E is true and A is false requires more explanation than the circumstance in which E and A are both true.[66]

At first sight, the distinction between normic support and evidential support looks much like the distinction between law-like statements and mere statistical generalizations in philosophy of science. Smith indeed makes the comparison:

> The distinction between the E_1-A relationship and the E_2-A relationship might fruitfully be compared to the distinction between statistical generalisations and normic or *ceteris paribus* generalisations widely accepted in the philosophy of science ....[67]

On closer inspection, however, there seems to be a difference. The problem in philosophy of science is that we lack a criterion for determining whether a particular sequence is law-like or merely accidental. All sequences that we encounter are finite, and we never know for sure whether or how they will continue — such are the lessons of Hume and Goodman. There might come a time when the sun does not rise, or rises only on Sundays, or rises in a completely random and unpredictable way. If, *per impossibile*, we knew for sure that a particular statement is a law-like statement, then we would be done; we could then safely use this knowledge in our predictions (which would hardly be predictions any more). And if, on the other hand, we knew that we are dealing with a mere statistical generalization, then we would realize that we should proceed cautiously, since we would find ourselves on rocky and unreliable ground.

The problem of law-like versus accidental generalizations is that *we have no way to determine* whether we are in the one or in the other situation, since

---

[66] Smith 2016, 40. The notion of normic support is further clarified in terms of normal worlds: "E normically supports a proposition A just in case A is true in all the most normal worlds in which E is true. Alternatively, we might say that E normically supports A just in case there is a world in which E is true and A is true which is more normal than any world at which E is true and A is false" (ibid., 42). Smith works out the technical details of normal worlds using not only David Lewis's method of nearby possible worlds, but also Wolfgang Spohn's ranking theory (Spohn 2012). See footnote 22 for Goldman on normal worlds.

[67] Smith 2010, 16. *Cf.* Smith 2016, 39-40, 128.

2.5 Smith's Normic Support

we do not know how the sequence will behave in the long run. This problem, however, no longer exists in the story about Bruce, for there we *know* whether we are dealing with normic or with evidential support. The reason that we have this knowledge is that we are *being told* how the sequence looks in the long run:

> In believing that [my] laptop is displaying a blue background, [Bruce is] actually running a *higher* risk of error that I would be in believing the same thing ... If this set-up were replicated again and again then, *in the long run*, [Bruce's] belief about [my] laptop would turn out to be false more often than my belief about my laptop.[68]

If a problem remains, then it is a different one. It is that the normic support that E_2 gives to A is in fact lower than the evidential support that E_1 gives to A. In other words, the inequality $P(A|E_1) > P(A|E_2)$ is not just apparent, based on a finite sequence of observations, but it persists in the long run — moreover we know that it does. Nonetheless, Smith advises us to base our belief in A on E_2 rather than on E_1. For only the E_2-A relationship is a relationship of normic support, since only E_2 can, according to Smith, be said to justify A.

We must confess that we have difficulty understanding this. Why base our belief on evidence which is less effective? Why rely on someone's perceptual information if we know that his eyesight is poor and our own information is more reliable? Smith's answer will be that in this case the poorer and less effective evidence is normically stronger. But what good is normic support that contradicts evidential support, not just now but also in the future? What is the sense of normic support as part of epistemic justification when it fundamentally disagrees with evidence we know to be true in the long run?

Of course, it might happen that we do not understand *why* $P(A|E_1)$ is greater than $P(A|E_2)$. But if we know that it *is* greater, should we not take that fact seriously? And does 'taking seriously' not mean that we act on E_1 rather than E_2? Note that if we do *not* act on E_1 in this case, a merciless opponent could use us as a money pump. And the fact that the rune of normic support is flaunted on our fluttering banner will not prevent us from becoming paupers in the fullness of time.

Smith himself makes no attempt to downplay these qualms:

> It may be rather tempting, however, for one to simply disregard such judgements [i.e., to trust E_2 rather than E_1] as confused or naïve. Perhaps we are

---

[68] Smith 2016, 35. our emphasis. We have adapted the example so that it fits the example that Smith describes in Smith 2010.

simply accustomed to relying upon perception when it comes to such matters and suspending any scruples about its fallibility. Once we do reflect carefully upon the fallibility of perception, so this thought goes, these troublesome intuitions are exposed as a kind of groundless prejudice. I'm not entirely convinced that this is the wrong thing to say — but I strongly suspect that it is.[69]

Note that we are not suggesting that justification is the same as evidential support. We agree with Smith that it is not, and Smith's many examples are convincing illustrations of this standpoint.[70] Let us be quite clear: we are not defending the risk minimization picture of justification. We do not think that justification can be defined as evidential support exceeding a certain threshold, nor are we saying that A is more justified by E_1 than by E_2 if the former gives more evidential support to A than the latter. Our claim is much weaker. We only maintain that probabilistic support, understood as the condition (2.1) and not to be confused with evidential support (see footnote 59), is a necessary condition of justification.

Probabilistic support is however not sufficient. Something has to be added to probabilistic support in order to turn it into justification. What is this 'something'? We do not know, but Smith thinks it is 'normalcy', that is the property that the support is normic. According to what he calls "the normic theory of justification", normalcy is necessary and sufficient for justification, but according to "the hybrid theory", it is only necessary.[71] It is not entirely clear which of the two theories Smith would finally choose; but as we have

---

[69] Smith 2016, 36.

[70] Some examples are about cases in which $P(A|B) > P(C|D)$, where all four variables are different. For instance, A = 'I will not win the lottery', B = 'I have bought one ticket in a fair lottery with a million tickets', C = 'The person in front of me will not suddenly drop dead', D = 'The person in front of me is young and healthy'. Then, even if $P(A|B) > P(C|D)$, it is still true, according to Smith, that D normically supports C, while B only makes A very likely (Smith 2010, 23). We believe that these cases do not provide much insight into the concept of justification, since they typically involve a comparison between two totally different domains.

Here is another example by Smith (from his talk 'When does evidence suffice for conviction?' on 30 April 2014 in Groningen; cf. Cohen 1977 and Nesson 1979). Imagine a hundred people walking out from an electronics store, each carrying a television. As it turns out, only one television has been paid for, so ninety-nine were stolen. Since Joe was one of the hundred, the probability that he is a thief is 0.99. Are we justified to believe that Joe is a thief? Not so, says Smith. Coos Engelsma proposed that epistemically we are justified, but not morally (private communication). Engelsma might have a point here, although this does not help the judge, who still has to weigh epistemic and moral justification against one another.

[71] Smith 2016, 76-79.

seen above there are cases in which normic support is inconsistent with evidential support. Hence we cannot have both as necessary conditions. What about probabilistic support? Can that be combined with normalcy? In fact it is possible, along the lines that Smith describes, to find cases where normic support also clashes with probabilistic support. Indeed, the example of Bruce and the computer can be tweaked to yield such a clash in the following way.

Suppose that it is I, and not Bruce, who sometimes observes my computer's screen. An evil hypnotist has however caused me to forget about the random generation of the background colour whenever I actually do observe the screen, but to remember how I programmed the boot routine when I am not looking at the screen. Now E_2 (the proposition that I see the colour to be blue) is true iff $\neg E_1$ is true, the proposition that I do not know about the random generator. In repeated boot sequences, $P(A|E_1) > P(A|E_2)$ becomes $P(A|\neg E_2) > P(A|E_2)$. Since it is E_2 that gives normic support to A, we have thereby constructed an inconsistency between normic and probabilistic support. Other less far-fetched examples are doubtless possible. And if probabilistic support and normic support *à la* Smith are not consistent, one or the other has to be rejected. The foregoing has made clear where our allegiance lies: we believe that it does not make sense to say that E justifies A without assuming that $P(A|E) > P(A|\neg E)$. Thus, when in the rest of this book we talk about 'justification' or 'epistemic justification' or 'probabilistic justification', we will always mean 'probabilistic support plus something else'. The indispensable rôle of probabilistic support as the inequality (2.1) will again become clear in Section 5.3, where we propose our view of justification as a trade-off.

2.6 Alston's Epistemic Probability

The second objection to our view is based on arguments by William Alston. As Alston sees it, Kolmogorov probability falls short of analyzing crucial epistemological issues. Although he in no way wishes to make light of the importance of probability, he believes that the Kolmogorovian rendering of it is deficient, not of course for understanding justification (since, according to him, there is no such thing), but as an instrument for analyzing epistemic desiderata. Especially the desiderata that Alston deems to be "the most fundamental ones", namely the so-called truth-conducive desiderata, rely heavily on concepts that have to do with probability.[72]

Alston lists five desiderata that are truth conducive, of which the first two are especially interesting for us:

1. The subject has adequate evidence (reasons, grounds ...) for the belief (A_i).
2. A_i is based on adequate evidence (reasons, grounds ...).[73]

The idea is that 1 is primarily about logical relations between propositions, while 2 is about the basing of beliefs. Alston deems 2 epistemically more desirable than 1, because he sees 2 as the actualization of the possibility provided by 1, and "the possibility of something desirable is less desirable than its realization."[74] He further notes that one could think of the basing relation as being causal in character, as long as one realizes that it is a special kind of causality, namely "the kind involved in the operation of input-output mechanisms that form and sustain, and so on, beliefs in a way that is characteristic of human beings."[75]

What does the term 'adequate' in 1 and 2 mean? Alston states that, if 1 and 2 are to be epistemic desiderata, "*adequacy* must be so construed that adequate evidence ... for A_i entails *the probable truth* of A_i."[76] In a further attempt to explain the meaning of 'adequate' he writes:

> The initial intuitive idea is that the ground is an indication that the belief is true, not necessarily a conclusive indication for that, but at least something that provides significant support for taking it to be true. Thus it is natural to think of an adequate ground of a belief A_i as something such that basing A_i on it confers a significant probability of truth on A_i.[77]

Alston stresses — and this is the salient point here — that his use of the word 'probability' deviates from the word as it occurs in a Kolmogorovian context. Probability for Alston is, as he dubs it, 'epistemic conditional probability' or for short 'epistemic probability'. It is subject to three constraints: it applies to beliefs, it is to be understood as 'the probability that a belief is true', and it

---

[72] Alston 2005a, 81.
[73] Ibid., 43, 81. Here and elsewhere we have replaced Alston's B by A_i.
[74] Ibid., 90.
[75] Ibid., 84.
[76] Ibid., 43. We have added the last italics.
[77] Ibid., 94.

2.6 Alston's Epistemic Probability

is essentially conditional.[78] As such these three constraints do not yet seem to carve out a non-Kolmogorovian probability, but Alston highlights three ways in which his epistemic conditional probability "fails to coincide with conditional probability as typically treated in probability theory."[79] Below we will discuss all three of them, arguing that none of them conflicts with orthodox probability theory.

The first way in which Alston's epistemic probability is supposed to deviate from standard probability theory hinges on the difference between doxastic and nondoxastic (primarily experiential) grounds of belief:

> First look at doxastic grounds. Suppose S's belief that *Susie is planning to leave her husband* (A_i) is based on S's belief *that Susie told her close friend, Joy, that she was* (A_j). To decide how strong an indication the belief that A_j is of the truth of the belief that A_i, we have to look at two things. First, if we stick for the moment as long as possible with the treatment in terms of propositions, the relation between the propositions that are the contents of these beliefs, A_j and A_i, is one factor that influences the conditional probability of A_i on A_j. But, second, we have to look at the epistemic status of the belief that A_j. For even if the conditional probability of A_i on A_j is high, that won't put S in a strong epistemic position in believing that of A_i if S has no good reason, or not a good enough reason, to believe that A_j. This consideration is sufficient to show that where the ground is doxastic the adequacy of the ground is not identical with the conditional probability of the propositional content of the target belief on the propositional content of the grounding belief.
>
> With nondoxastic grounds, on the other hand, we are not faced with this second factor. Where my ground is a certain visual experience rather than, for example, a belief that I have that experience, the ground is a fact rather than a belief in a fact. Hence no problem can arise with respect to the epistemic status of the ground since that ground is not the sort of thing that can have an epistemic status. And so the adequacy of a nondoxastic ground coincides exactly with the conditional probability of the propositional content of the belief in that fact, construed as a true proposition. Here conditional probability as treated in probability theory can translate directly into an epistemic status.[80]

Here Alston suggests that, if the ground A_j is a belief, then standard probability theory will look only at the conditional probability of the target belief

---

[78] Ibid., 95. Alston maintains that "conditional probabilities are in the center of the picture for the epistemology of belief" (ibid.). We fully agree with him here. After all, our condition of probabilistic support is made up of conditional probabilities, and our view that epistemic justification is intrinsically relational (see Section 2.2) also accommodates that point.
[79] Ibid., 97.
[80] Ibid.

A_i, given the grounding belief A_j:

$$P(A_i|A_j).$$

In contrast, his own epistemic probability also accounts for the epistemic status of A_j itself, $P(A_j)$, so that we would obtain

$$P(A_i|A_j)P(A_j).$$

Alston's suggestion is however incorrect. Standard probability accounts not only for the epistemic status of A_j, namely $P(A_j)$, but also for the epistemic status of $\neg A_j$, namely $P(\neg A_j)$. The latter is as important as the former. For if the probability of A_i is conditioned on the probability of A_j, then one can only calculate the former probability if one also takes into account what that probability would be in case A_j is false. In standard probability theory the fact that the probability of A_i is conditioned by the probability of A_j is expressed by the rule or law of total probability:

$$P(A_i) = P(A_i|A_j)P(A_j) + P(A_i|\neg A_j)P(\neg A_j). \tag{2.4}$$

So it seems that Alston's mistake is twofold. First he incorrectly suggests that standard probability theory does not consider $P(A_j)$, and second he himself neglects the relevance of the second term in (2.4). As we will see in the next chapter, the erroneous neglect of the second term of the rule of total probability is a mistake that has occurred more often in philosophy; even such notable scholars as Clarence Irving Lewis and Bertrand Russell fell prey to it.

Formula (2.4) also enables us to understand better what Alston says about the case in which the ground A_j is not a belief. Alston correctly notes that, if the ground is nondoxastic, it is a fact rather than a belief in a fact. This fact, "construed as a true proposition", has probability 1, so $P(A_j) = 1$. This means that $P(\neg A_j) = 0$, and thus that (2.4) reduces to the "conditional probability as treated in probability theory", viz.

$$P(A_i|A_j),$$

in accordance with what Alston states.

The second respect in which Alston's epistemic probability allegedly fails to conform to the standard probability calculus concerns the relevance that the ground A_j has to the target A_i. Alston illustrates his point by referring to the case where A_i is a necessary truth or necessary falsehood:

2.6 Alston's Epistemic Probability

> In the standard probability calculus the probability of every necessary truth is 1 and the probability of every necessary falsehood is 0. This makes it impossible to use conditional probabilities in assessing the adequacy of grounds for necessarily true or false beliefs. Since every necessary truth has a probability of 1, no matter what else is the case, its conditional probability on any proposition whatever is 1. This 'rigidity' of the probability of necessary truths prevents it from capturing what we are after in thinking of the adequacy of the grounds. One who supposes that a person who believes $2 + 2 = 4$ on the basis of the belief that all crows are black, thereby believes the former on the basis of a significantly adequate ground, is missing the epistemological boat. In thinking of a ground as adequate to some considerable degree, we take it to *render* what it grounds as more or less probable. It must make a significant difference to the probability of the grounded belief. ...The axioms of arithmetic are adequate grounds for $2 + 2 = 4$, unlike the proposition that all crows are black. But this will have to be explained on the basis of some other than the probability calculus.[81]

In this passage, Alston is criticizing the standard probability calculus on two points. First, in order for the ground A_j to be adequate (or relevant) for the target A_i, A_j must render A_i more or less probable. But if A_i is a necessary truth or falsehood, then standard probability implies that A_j will not render A_i more or less probable. Thus A_j will not be adequate or relevant to A_i, and this is counterintuitive. Second, *whether* A_j renders A_i more or less probable, and thus whether A_j is (ir)relevant to A_i, will have to be determined outside the probability calculus, and this is troublesome.

What to make of these two points? Let's begin with the first one. We will follow Alston in describing an adequate or relevant ground as one that makes the target more of less probable; this in fact sits well with our condition of probabilistic support. Alston is right that in standard probability theory, if the target A_i is a tautology or a contradiction, then the ground A_j will not render A_i more or less probable, and will in that sense be irrelevant to the target. But why call this counterintuitive or otherwise problematic? It would be stranger, and even inconsistent, if something that already had the maximum probability value would acquire an even higher one.

Moreover, calling A_i a tautology or a contradiction (necessary truth or falsehood) already presupposes a system *in which* A_i has that character. That is, if A_i is $2 + 2 = 4$, then A_i is a tautology relative to any system equivalent to the axioms of arithmetic. So if A_j is such a system, then $P(A_i|A_j) = 1$. The situation remains the same if we add to A_j the proposition that all crows are black: $P(A_i|A_j$ and all crows are black$) = 1$. What if the ground A_j consists

---

[81] Ibid., 97-98.

only of the proposition that all crows are black and nothing more, in particular nothing that is equivalent to the axioms of arithmetic? In that case A_i and A_j are independent of each another,

$$P(A_i \wedge A_j) = P(A_i)P(A_j),$$

which implies that $P(A_i|A_j)$ is equal to $P(A_i)$. Here A_j is also irrelevant to A_i (and vice versa), since A_j does not render A_i more or less probable. However, A_j is now irrelevant for a completely different reason: it is irrelevant, not because it already confers upon A_i the maximum probability value, but because it is independent of A_i.

Alston's second point of criticism is that we have to go outside probability theory in order to establish whether a ground is relevant to a target: we can only explain that Peano arithmetic is relevant to $2 + 2 = 4$, and that 'crows are black' is not, on some other basis than the probability calculus. We think there is a confusion here. A comparison with standard logic might help. Ask yourself: is A_j relevant to A_i in the sense that A_j makes A_i *true* (rather than probable)? The answer to this question depends first and foremost on what A_j and A_i *mean*. Let A_i mean 'Feike can swim'.[82] If A_j means 'Feike is a Frisian and all Frisians can swim', then A_j is clearly relevant to A_i, and 'if A_j then A_i' expresses a logical connection. The situation is the same in standard probability theory. If A_j means 'Feike is a Frisian and 9 out of 10 Frisians can swim', then A_j is obviously relevant to A_i, and $P(A_i|A_j) = 0.9$ expresses a logical connection. Of course, whether A_j is true can only be established outside logic or probability theory: we need empirical information in order to determine whether all Frisians, or only 9 out of 10, can swim. The fact that both in logic and in probability theory we often need the world to determine whether premises are true is a fact of life, it is a deficiency neither of logic nor of probability theory.

The third way in which, according to Alston, epistemic conditional probability fails to coincide with conditional probability, as treated in probability theory, has to do with the basing relation. Whereas ordinary conditional probability is typically concerned with relations between propositional contents, Alston's conditional probability focuses on relations between beliefs and their grounds; in contrast to the former, the latter are basing relations. In the previous pages we have argued that this difference between the two relations can be seen as a difference between interpretations of the probability calculus: logical or conceptual in the one case, and nomological or causal in

---

[82] We borrow the example from Alston, who in turn has borrowed it from Alvin Plantinga (ibid., 105).

2.6 Alston's Epistemic Probability

the other. This would mean that, for Alston, A_j is the ground on which A_i is (probabilistically) based. However, this appears to be too simple, and not quite in accordance with what Alston writes. Alston does not say that the belief A_i is based on a ground, but that:

> the basing of the belief $[A_i]$ on the ground ... is the condition on which the probability of the belief $[A_i]$ is conditional.[83]

So rather than saying that the condition A_j *is* the ground of A_i, Alston appears to say that A_j is the condition of *being based on a ground*. However, even that reconstruction might not be what Alston has in mind, for he writes:

> ... that on which the probability of the target belief, A_i, is conditional differs in the two cases [the case of Alston's conditional probability and that of ordinary conditional probability]. For the latter, it is the conditioning propositions, taken as true. For the former, it is the basing of A_i on a ground of a certain degree of adequacy.[84]

Thus A_j is the condition of being based *with a certain degree of adequacy* on a ground. Alston continues:

> And that degree of adequacy is a function of more than the relation of propositional contents. As we have seen, it is also a function of the epistemic status of any beliefs in the ground. So in addition to the difference between a proposition-proposition(s) relationship and a belief-ground relationship, even the factors relevant to the status of the conditioning item(s) do not exactly match.[85]

Here Alston suggests that the difference between the two kinds of conditional probability, the traditional and the Alstonian one, is that only the latter accounts for $P(A_j)$. However, we have seen that this is not so. When A_i is conditioned on A_j, traditional probability theory gives the probability of A_i via the rule of total probability (2.4), and (2.4) clearly also accounts for $P(A_j)$.

We conclude that Alston's concept of epistemic conditional probability actually coincides with the concept of conditional probability as it is treated in traditional probability theory. None of the three alleged differences that Alston describes constitutes a departure from Kolmogorovian orthodoxy.

---

[83] Ibid., 98.
[84] Ibid., 99.
[85] Ibid.

Open Access This chapter is licensed under the terms of the Creative Commons Attribution 4.0 International License (http://creativecommons.org/licenses/by/4.0/), which permits use, sharing, adaptation, distribution and reproduction in any medium or format, as long as you give appropriate credit to the original author(s) and the source, provide a link to the Creative Commons license and indicate if changes were made.

The images or other third party material in this chapter are included in the chapter's Creative Commons license, unless indicated otherwise in a credit line to the material. If material is not included in the chapter's Creative Commons license and your intended use is not permitted by statutory regulation or exceeds the permitted use, you will need to obtain permission directly from the copyright holder.

Chapter 3
The Probabilistic Regress

Abstract
During more than twenty years Clarence Irving Lewis and Hans Reichenbach pursued an unresolved debate that is relevant to the question of whether infinite epistemic chains make sense. Lewis, the nay-sayer, held that any probability statement presupposes a certainty, but Reichenbach profoundly disagreed. We present an example of a benign probabilistic regress, thus showing that Reichenbach was right. While in general one lacks a criterion for distinguishing a benign from a vicious regress, in the case of probabilistic regresses the watershed can be precisely delineated. The vast majority ('the usual class') is benign, while its complement ('the exceptional class') is vicious.

3.1 A New Twist

The previous chapter indicated how intricate the debate about epistemic justification has become. A mixed bag of knotty details and drawbacks complicates the subject, giving rise to a variety of different positions. But although nobody knows what exactly epistemic justification is, the idea that it involves probabilistic support is widespread among epistemologists of all sorts and conditions. Internalists, externalists, foundationalists, anti-foundationalists, evidentialists and reliabilists: most of them assume that 'A_j justifies A_i' implies that A_i somehow receives probabilistic support from A_j.

In this chapter and the ones to follow we want to make clear how significant this turn towards probability actually is, and what surprising consequences it has. The debate about epistemic regresses acquires a completely new twist when Kolmogorovian probability is brought into the picture; for as we will see a probabilistic regress turns out to be immune to many of the objections that have routinely been raised against the traditional regress of entailments. The situation is to a certain extent reminiscent of the two causal regresses that we encountered in Chapter 1. Whereas a causal series *per se* only makes sense if it has a first member, this is not so for a causal series *per accidens*. Similarly, as we will argue, a traditional regress of entailments needs a first member, but a regress of probabilistic support may not.

In the present chapter we will describe the concept of a probabilistic regress, that is a regress in which (1.1) of Chapter 1,

$$q \longleftarrow A_1 \longleftarrow A_2 \longleftarrow A_3 \longleftarrow A_4 \ldots$$

is reinterpreted as: q is probabilistically supported by A_1, which is probabilistically supported by A_2, and so on, *ad infinitum*.[1] It is assumed that every link in this chain satisfies the condition of probabilistic support (2.1). As we have seen, this condition is quite weak, falling considerably short of the title 'justification'. But for our purposes this minimal requirement is enough.

Our exposition of a probabilistic regress takes as its starting point a historical debate between Hans Reichenbach (1891-1953) and Clarence Irving Lewis (1883-1964). Lewis and Reichenbach are both early defenders of the view that epistemic justification is probabilistic in character, holding that A_j might justify A_i even if the former does not logically entail the latter but only provides probabilistic support. They disagree, however, as to the implications of this claim. Lewis insists that probabilistic justification must spring from a ground that is certain, whereas Reichenbach maintains that probabilistic justification remains coherent, even if it is not rooted in firm ground. The disagreement between Lewis and Reichenbach extended over more than two decades, from 1930 until 1952, and it is well documented in letters and in journal contributions.

In Sections 3.2 and 3.3 we will give an overview of the dispute. We first describe Lewis's main claim, viz. that any proposition of the form 'q is probable' or 'q is made probable by A_1' must presuppose a proposition that is certain. Lewis's argument for this claim is that without such a presupposition we will end up with a probabilistic regress that has the absurd consequence of always yielding probability value zero for q. Next we describe Reichenbach's objection to this argument. We then explain that Lewis is not convinced by it and challenges Reichenbach to produce a counterexample,

---

[1] The term 'probabilistic regress' was coined by Frederik Herzberg (Herzberg 2010).

i.e. a probabilistic regress that yields a number other than zero for the target proposition q.

Reichenbach never took up Lewis's challenge, but we will meet it in Section 3.4. By presenting a probabilistic regress that converges to a non-zero limit, we demonstrate that a target can have a definite and computable value, even if it is probabilistically justified by a series that continues *ad infinitum*. In this manner we show that Reichenbach rather than Lewis was correct, and also that a probabilistic regress can make sense.

The counterexample to Lewis in Section 3.4 has a simple, uniform structure. In Section 3.5 we offer a nonuniform and thus more general counterexample. Both counterexamples belong to what we call 'the usual class', i.e. the class of probabilistic regresses that yield a well-defined probability for the target proposition. We distinguish it from 'the exceptional class', which contains the probabilistic regresses that are not well-defined. In Section 3.6 we will spell out the conditions for membership of the usual and the exceptional classes. As it turns out, exceptional probabilistic regresses are characterized by the fact that here probabilistic support comes very close to entailment. Not surprisingly, therefore, probabilistic regresses in the exceptional class need a ground in order to bestow a value on the target, and in that sense count as vicious.

The uniform and the nonuniform counterexamples in 3.4 and 3.5 are rather abstract in nature; but in Section 3.7 we offer two real-life probabilistic regresses, based on the development of bacteria.

3.2 The Lewis-Reichenbach Dispute

In 1929 Lewis published his first major work, *Mind and the world order. An outline of a theory of knowledge*.[2] Here he starts from the traditional view that our knowledge is partly mathematical and partly empirical. The mathematical part deals with knowledge that is *a priori* and analytic; the empirical part concerns our knowledge of nature. This knowledge of nature, says Lewis, is always only probable:

> ...all empirical knowledge is probable only ...our knowledge of nature is a knowledge of probabilities.[3]

---

[2] The present section is based on Peijnenburg and Atkinson 2011.
[3] Lewis 1929, 309-310.

Since the crucial issue for any theory of knowledge is the character of empirical knowledge, it follows that

> ...the problem of our knowledge ... is that of the validity of our probability judgements.[4]

What about the validity of probability statements? In *Mind and the world order*, Lewis stresses time and time again that probability judgements only make sense if they are based on something that is certain:

> The validity of probability judgements rests upon ...truths which must be certain.[5]

> ...the immediate premises are, very likely, themselves only probable, and perhaps in turn based upon premises only probable. Unless this backward-leading chain comes to rest finally in certainty, no probability-judgment can be valid at all.[6]

Lewis is not the only philosopher who has argued that probability judgements presuppose certainties. The idea can already be found in David Hume's *Treatise of human nature* and it has also been defended by, among others, Keith Lehrer, Richard Fumerton, and Nicholas Rescher.[7] Lewis is however one of the few who discusses the claim in more detail. His explanation can be summarized as follows.

A statement of the form 'q is probable' or 'the probability of q is x' is in fact elliptical for 'q is probable, given A_1', or 'the probability of q given A_1 is x', where x is a number between one and zero. In symbols: the unconditional $P(q) = x$ is elliptical for the conditional $P(q|A_1) = x$. In many cases, A_1 is itself only probable, so we obtain 'A_1 is probable', which is shorthand for 'A_1 is probable, given A_2'. Again, if A_2 is only probable, we need A_3, et cetera. A probabilistic regress threatens. Lewis's claim is that in the end we must encounter a statement, p, that is certain (or has probability 1 — we will not distinguish here between these two cases):

$$q \longleftarrow A_1 \longleftarrow A_2 \longleftarrow A_3 \longleftarrow A_4 \longleftarrow \ldots \longleftarrow p.$$

Denying that this is so, and claiming that such a certain p is not needed, says Lewis, amounts to making nonsense of the original statement ('q is

---

[4] Ibid., 308.
[5] Ibid., 311.
[6] Ibid., 328-329.
[7] Hume 1738/1961, 178; Lehrer 1974, 143; Fumerton 2004, 162; Fumerton and Hasan 2010; Rescher 2010, 36-37.

3.2 The Lewis-Reichenbach Dispute

probable') itself. Thus we can only give a probability value to a target, q, if we suppose that there is a ground or foundation, p, that is certain.[8].

Reichenbach read *Mind and the world order* soon after it came out. Although he concurred with many of Lewis's reasonings, he profoundly disagreed with the claim that probability statements only make sense if they are based on certainties. On July 29, 1930, he sent Lewis a letter, enclosing some of his own manuscripts. Unfortunately this letter is now lost. We only know of its existence from a reply that Lewis wrote to Reichenbach, dated August 26, 1930.[9] We are unable to infer from this reply what exactly Reichenbach had written, since Lewis mainly writes about the manuscripts that Reichenbach had sent him.[10]

Between 1930 and 1940 a correspondence developed, which was partly about practical matters (Reichenbach had fled Berlin in 1933 and went to Istanbul, from where he tried to find an academic position in the U.S.A.), and partly about Lewis's claim that probability judgements presuppose certainties. As far as the latter is concerned, it is clear that Reichenbach's arguments did not convince Lewis, for sixteen years later, in his book *An analysis of knowledge and valuation*, Lewis stresses the same point again:

> If anything is to be probable, then something must be certain. The data which themselves support a genuine probability, must themselves be certainties.[11]

The disagreement between Lewis and Reichenbach reached its height in December 1951, at the forty-eighth meeting of the Eastern Division of the American Philosophical Association at Bryn Mawr. At that meeting there was a symposium on 'The Given', where Lewis, Reichenbach and Nelson Goodman read papers. Their contributions were published a year later in *The Philosophical Review*, and there we learn that Lewis sticks to his guns:

---

[8] As James Van Cleve has noted, Lewis's text appears to be ambiguous between two readings (Van Cleve 1977, 323-324). According to the first, Lewis says something like: 'The probability of q given p is x, and moreover p is certain'. In symbols: $P(q|p) = x$ and $P(p) = 1$. According to the second reading he says: 'It is certain that the probability of q given p is x', that is $P(P(q|p) = x) = 1$. It can however be proven that the two readings are equivalent, so this ambiguity is merely apparent. We will come back to this matter in Chapter 7.

[9] "Your very kind letter of July 29th has reached me, here at my summer address." The summer address was, by the way, Briar Hill in New Hampshire, close to Vermont.

[10] And apparently did not know quite what to do with them: "I find difficulty in understanding the ground from which they arise."

[11] Lewis 1946, 186.

> The supposition that the probability of anything whatever always depends on something else which is only probable itself, is flatly incompatible with the assignment of any probability at all.[12]

But Reichenbach, too, insisted on his own views. Already in his major epistemological work, *Experience and prediction*, he had found an apt metaphor for his anti-foundationalist position:

> All we have is an elastic net of probability relations, floating in open space.[13]

Fifteen years later Reichenbach still had the same conviction. He calls the claim of Lewis that probabilities must be grounded in certainties "just one of those fallacies in which probability theory is so rich".[14] In an attempt to understand the root of the fallacy he writes:

> We argue: if events are merely probable, the statement about their probability must be certain, because ... Because of what? I think there is tacitly a conception involved according to which knowledge is to be identified with certainty, and probable knowledge appears tolerable only if it is embedded in a framework of certainty. This is a remnant of rationalism.[15]

And being a rationalist would of course be a thorn in Reichenbach's logical-empiricist side. Lewis, in turn, rejects the accusation of being an old fashioned rationalist and replies that, on the contrary, he is trying to save empiricism from what he calls 'a modernized coherence theory' like that of his opponent. He writes:

> ...the probabilistic conception [of Reichenbach] strikes me as supposing that if enough probabilities can be got to lean against one another they can all be made to stand up. I suggest that, on the contrary, unless some of them can stand up alone, they will all fall flat.[16]

Who is right in this debate? Some authors, such as James Van Cleve and Richard Legum, have argued that it is Lewis.[17] To explain why we dissent, we will first spell out the argument that Lewis puts forward in support of his claim that probability judgements presuppose certainties. It is true that the negation of Lewis's claim leads to an infinite regress, but since not all regresses are vicious, an argument is required in order to show that this particular regress is of the unacceptable kind.

---

[12] Lewis 1952, 173.
[13] Reichenbach 1938, 192.
[14] Reichenbach 1952, 152.
[15] Ibid.
[16] Lewis 1952, 173.
[17] Van Cleve 1977; Legum 1980.

3.3 Lewis's Argument

As Mark Pastin correctly notes, the claim that probabilities presuppose certainties was repeated by Lewis "throughout his writings but [he] gave little attention to defending it".[18] The most extensive defence can be found in *Mind and the world order*, which contains the following argument:

> Nearly all the accepted probabilities rest upon more complex evidence than the usual formulations suggest; what are accepted as premises are themselves not certain but highly probable. Thus our probability judgement, if made explicit, would take the form: the probability that A is B is a/b, because if C is D, then the probability that A is B is m/n, and the probability of 'C is D' is c/d (where $m/n \times c/d = a/b$). But this compound character of probable judgement offers no theoretical difficulty for their validity, provided only that the probability of the premises, when pushed back to what is more and more ultimate, somewhere comes to rest in something certain.[19]

In other words, Lewis says that the judgement

$$A \text{ is } B \quad \text{is probable}, \tag{3.1}$$

is elliptical for

$$A \text{ is } B \quad \text{is probable, given } C \text{ is } D. \tag{3.2}$$

Since we are dealing with empirical knowledge, *C is D* is itself also only probable. The judgement '*C is D* is probable' is in turn elliptical for '*C is D* is probable, given *E is F*'. And so on.

We can formalize and quantify (3.1) and (3.2) by

$$P(A \text{ is } B) = a/b \tag{3.3}$$

which is elliptical for

$$\begin{aligned} P(A \text{ is } B) &= P(A \text{ is } B | C \text{ is } D) \times P(C \text{ is } D) \\ &= m/n \times c/d \\ &= a/b, \end{aligned} \tag{3.4}$$

where a/b, m/n and c/d are probability values between 1 and 0. Now of course the probability that *C is D* may also be elliptical. If this series were to

---

[18] Pastin 1975, 410.
[19] Lewis 1929, 327-28. Here 'A is B' means something like 'all A-things are B-things'. We have replaced Lewis's 'P is Q' and 'p/q' by 'C is D' and 'c/d'.

go on and on, then, because all the factors in the multiplication are probabilities and thus positive numbers less than one, the probability of the original proposition *A is B* would always tend to zero. But this is ridiculous, so the series of probability judgements must come to a stop in a statement that is certain. This is Lewis's argument for his claim that bestowing a probability value on a target presupposes the acceptance of a ground that is certain: without such a ground, the probability of the target will go to zero.

Lewis's argument is however simply mistaken. For $P(A\ is\ B)$ is not elliptical for the product $P(A\ is\ B|C\ is\ D) \times P(C\ is\ D)$, but for the following sum of products:

$$P(A\ is\ B) = P(A\ is\ B|C\ is\ D) \times P(C\ is\ D) \\ + P(A\ is\ B|\neg(C\ is\ D)) \times P(\neg(C\ is\ D)). \quad (3.5)$$

The first term of (3.5) coincides with (3.4), but (3.5) contains a second term, which Lewis forgets. He ignores the fact that, if the probability of *A is B* is conditioned by the probability of *C is D*, then you can only calculate the former probability if you also take into account what that probability is in case *C is D* is false.[20] Eq.(3.5) is an instance of the rule of total probability, which is a theorem of the calculus that Andrey Kolmogorov developed in his *Grundbegriffe der Wahrscheinlichkeitsrechnung*.

Kolmogorov published his *Grundbegriffe* in 1933, which might explain Lewis's mistake. The same can however not be said of Bertrand Russell. In 1948, nineteen years after *Mind and the world-order*, Russell published *Human knowledge: its scope and limits*. Part 5 of this book is devoted to the concept of probability, and there Russell criticizes several theories of probability, including Reichenbach's theory in his *Wahrscheinlichkeitslehre* of 1935. It is interesting that, quite independently of Lewis (for he does not mention him anywhere), Russell claims that attributing a probability value to a proposition presupposes a certainty. Moreover, he defends this claim with the same erroneous argument that Lewis had used. Russell writes:

> At the first level, we say that the probability that an *A* will be a *B* is m_1/n_1; at the second level, we assign to this statement a probability m_2/n_2, by making it one of some series of similar statements; at the third level, we assign a

---

[20] Mark Pastin seems to interpret Lewis as talking about the probability of the *conjunction* of the propositions '*A is B*' and '*C is D*' (Pastin 1975, 413). In this reading, Eq.(3.5) would be replaced by $P((A\ is\ B)\ and\ (C\ is\ D)) = P(A\ is\ B|C\ is\ D) \times P(C\ is\ D)$, and this expression has no second term. However in this case there would not be a justificatory chain in which one proposition justifies the other. See footnote 31.

3.3 Lewis's Argument

probability m_3/n_3 to the statement that there is a probability m_2/n_2 in favour of our first probability m_1/n_1; and so we go on forever. If this endless regress could be carried out, the ultimate probability in favour of the rightness of our initial estimate m_1/n_1 would be an infinite product

$$m_2/n_2 \times m_3/n_3 \times m_4/n_4 \ldots$$

which *may be expected to be zero*.[21]

In other words, Russell argues that a series of statements like

$s_1 = A$ is B
$s_2 = $ The probability of s_1 is m_1/n_1
$s_3 = $ The probability of s_2 is m_2/n_2
...

implies that the probability of s_1 will tend to zero.[22] The argument is the same as that of Lewis: the probability of s_1 is the outcome of the multiplication of an infinite number of factors each of which is smaller than 1. It thus fails for precisely the same reason as does Lewis's argument. If a proposition

---

[21] Russell 1948, 434; our italics. Where Russell has α and β we have used A and B. It is assumed that $0 < m_i/n_i < 1$ for all i. Presumably Russell, a competent mathematician, wrote 'may be expected to be zero' because he knew that there exist infinite products of factors, all less than one, that converge (i.e. that yield well-defined, non-zero values). In this connection it is interesting that Quine, in his 1946 Lectures on David Hume's Philosophy (Quine 2008), indeed makes the point that such a product can be convergent: in fact he gives an explicit example. He fails, however, to note that the point is irrelevant, for the probabilities in question should not be multiplied together (because of the second term in (3.5)). Thanks to Sander Verhaegh for bringing Quine's lectures to our notice. We return to Quine's reasoning in Chapter 7.

[22] Note that Russell here speaks about higher-order probability statements rather than about the probability of a reference class in a conditional probability statement (see footnote 8 for the difference). Russell says that such a series of higher-order probability statements "leads (one is to suppose) to a limit-proposition, which alone we have a right to assert. But *I do not see how this limit-proposition is to be expressed*. The trouble is that, as regards all the members of the series before it, we have no reason ... to regard them as more likely to be true than to be false; they have, in fact, no probability that we can estimate." (Russell 1948, 435; our italics). In other words, Russell suggests that we cannot attribute a probability value to s_1 because we are unable to compute the limit of the series. This seems to be at odds with his earlier claim that the value of s_1 goes to zero, but we will not dwell on the matter here. In the next section we will rather specify the limit proposition that Russell was vainly trying to express.

with probability x is conditioned by a proposition with probability y, then the probability of the first proposition is not given by xy, as Russell says, but by $xy + x'(1-y)$, where x' is the probability that the first proposition is true if the second is false, and $(1-y)$ is the probability that the second proposition is indeed false. Just like Lewis, Russell forgets the second term in the rule of total probability, namely $x'(1-y)$.

Reichenbach notices that Russell makes the mistake, and points it out to him in a letter of March 28, 1949.[23] Russell clearly acknowledges his oversight, as we see from his reply three weeks later.[24] Lewis, on the other hand, seems to have persisted in his error, and Reichenbach confronts him with this fact in 1951, at the forty-eighth meeting of the American Philosophical Association at Bryn Mawr. Lewis appears however not to be impressed by Reichenbach's amendment:

> ...even if we accept the correction which Reichenbach urges here, I disbelieve that it will save his point. For that, I think he must prove that, where any regress of probability-values is involved, the progressively qualified fraction measuring the probability of the quaesitum will converge to some determinable value other than zero; and I question whether such a proof can be given.[25]

In other words, Lewis fails to see the relevance of the second term in (3.5): he simply does not believe that an infinite regress of probabilities can converge to some value other than zero. Even if we *do* take Reichenbach's amendment into account, Lewis still thinks that an infinite series of probability statements conditioned by probability statements will always converge to zero. And he defies Reichenbach to prove the contrary. As far as we know Reichenbach never took up the challenge. Perhaps he planned to, but never got around to it; or maybe he had difficulties finding what Russell called "the limit proposition" (see footnote 22); or perhaps he simply got tired of the debate. We will presumably never know, for in April 1953 Reichenbach died in California of a heart attack.

---

[23] The letter is printed in the volume with selected writings of Hans Reichenbach edited by Maria Reichenbach and Robert Cohen (Reichenbach and Cohen 1978, 405-411).

[24] "I perceive already that you are right as to the mathematical error that I committed on page 416" (letter from Russell to Reichenbach, April 22, 1949). Page 416 corresponds to page 434 in reprints of Russell's book. We are grateful to Mr. L. Lugar and Ms. B. Arden of the Pittsburgh Archive for sending us a copy of Russell's letter.

[25] Lewis 1952, 172.

3.4 A Counterexample

In the next section we will take up Lewis's gauntlet by presenting a counterexample to his argument that a "regress of probability-values" always tends to zero. This counterexample involves an infinite iteration of the rule of total probability. Although this iteration produces a much more complicated regress than the simple product that Russell and Lewis had envisaged, it leads to a perfectly well-defined, and moreover nonzero probability for the target proposition. It thus also produces the "limit-proposition" that Russell was looking for.[26]

3.4 A Counterexample

Let our target proposition q be probabilistically justified by proposition A_1. We have seen that the unconditional probability of q, namely $P(q)$, can be calculated from the rule of total probability:

$$P(q) = P(q|A_1)P(A_1) + P(q|\neg A_1)P(\neg A_1). \tag{3.6}$$

To make contact with Lewis's argument, we can take q to be 'A is B' and A_1 to be 'C is D'. If A_1 is probabilistically justified by A_2, then $P(A_1)$ can be calculated from another instance of the rule,

$$P(A_1) = P(A_1|A_2)P(A_2) + P(A_1|\neg A_2)P(\neg A_2), \tag{3.7}$$

and if A_2 is in turn probabilistically justified by A_3 we have to repeat the rule again,

---

[26] Dennis Dieks put forward the possibility that Lewis might have been interested only in those probabilistic regresses in which the second term may be legitimately ignored (Dieks 2015). Dieks' suggestion is intriguing, but it causes difficulties. First, why did not Lewis make this explicit? In his debate with Reichenbach there appear to have been opportunities enough. Second, even if A_{n+1} has been called a reason for A_n, we should not overlook the fact that other propositions, contained in the negation of A_{n+1}, can well contribute to the justification of A_n. As Johan van Benthem phrases it: "[$P(A_n|\neg A_{n+1})$] measures intuitively the 'bonus' that A_n receives even if A_{n+1} were untrue. This inclusion might perhaps sound odd if we have just introduced A_{n+1} as reason for A_n — but we may, neither here nor in argumentation generally, ignore the fact that the postulated claim can already enjoy support without A_{n+1}" (Van Benthem 2015, 148, our translation from the Dutch; cf. Peijnenburg 2015, 205-206). In any case, if Dieks were correct this would considerably restrict the domain in which the Lewisian approach could apply, and it would appear to be inconsistent with the probability calculus.

$$P(A_2) = P(A_2|A_3)P(A_3) + P(A_2|\neg A_3)P(\neg A_3). \tag{3.8}$$

Can we continue this repetition, thus allowing for propositions being probabilistically justified by other propositions, being probabilistically justified by still other propositions, *ad infinitum*? It might look as though we cannot. How would we ever be able to calculate $P(q)$ if it is the outcome of an infinite regress of instances of the rule of total probability? The calculation seems at first sight to be too lengthy and too complicated for us to complete. After all, insertion of Eq.(3.7), together with

$$P(\neg A_1) = P(\neg A_1|A_2)P(A_2) + P(\neg A_1|\neg A_2)P(\neg A_2) \tag{3.9}$$

into the right-hand side of Eq.(3.6) leads to an expression with four terms, namely:

$$P(q) = P(q|A_1)P(A_1|A_2)P(A_2) + P(q|\neg A_1)P(\neg A_1|A_2)P(A_2) + \tag{3.10}$$
$$P(q|A_1)P(A_1|\neg A_2)P(\neg A_2) + P(q|\neg A_1)P(\neg A_1|\neg A_2)P(\neg A_2).$$

A repetition of this manoeuvre to express $P(A_2)$ and $P(\neg A_2)$ in terms of $P(A_3)$ and $P(\neg A_3)$ would produce no less than eight terms. After $n+1$ steps, the number of steps is 2^{n+1}, yielding an ungainly expression that seems hard to evaluate in a simple, closed form.

There is however a way to reduce this complication of the rapidly increasing number of terms. In explaining this we first simplify the notation by abbreviating (3.6) by setting the two conditional probabilities, $P(q|A_1)$ and $P(q|\neg A_1)$, equal to α and β:

$$\alpha = P(q|A_1) \qquad \beta = P(q|\neg A_1). \tag{3.11}$$

Now $P(q)$ becomes:

$$P(q) = \alpha P(A_1) + \beta P(\neg A_1)$$
$$= \alpha P(A_1) + \beta[1 - P(A_1)]$$
$$= \beta + (\alpha - \beta)P(A_1). \tag{3.12}$$

Clearly, we can only compute $P(q)$ if we know $P(A_1)$. Of course, we also have to know the values of the conditional probabilities α and β. Their status is however rather different from that of the unconditional probabilities, and we will come back to this matter in detail in Chapter 4. At this juncture, we simply assume that α and β are given, and that they are the same from link to link (the latter assumption is dropped in the next section). But what *is* the value of $P(A_1)$? We do not know. However, we do know that A_1 is

3.4 A Counterexample

probabilistically justified by A_2, and so we can calculate $P(A_1)$ in terms of $P(A_2)$, and so on:

$$P(A_1) = \beta + (\alpha - \beta)P(A_2)$$
$$P(A_2) = \beta + (\alpha - \beta)P(A_3)$$
$$P(A_3) = \beta + (\alpha - \beta)P(A_4).$$

We can now see how to get rid of the unknown unconditional probabilities, namely by nesting the formulas. Thus we can remove $P(A_1)$ by substituting its value into (3.12), so that we obtain:

$$\begin{aligned} P(q) &= \beta + (\alpha - \beta)P(A_1) \\ &= \beta + (\alpha - \beta)[\beta + (\alpha - \beta)P(A_2)] \\ &= \beta + \beta(\alpha - \beta) + (\alpha - \beta)^2 P(A_2). \end{aligned} \quad (3.13)$$

Next, by inserting the value of $P(A_2)$ into (3.13) we attain

$$\begin{aligned} P(q) &= \beta + \beta(\alpha - \beta) + (\alpha - \beta)^2[\beta + (\alpha - \beta)P(A_3)] \\ &= \beta + \beta(\alpha - \beta) + \beta(\alpha - \beta)^2 + (\alpha - \beta)^3 P(A_3), \end{aligned} \quad (3.14)$$

by which we got rid of $P(A_2)$. And so on. After a finite number m of steps we obtain the following formula:

$$P(q) = \beta + \beta(\alpha - \beta) + \beta(\alpha - \beta)^2 + \ldots + \beta(\alpha - \beta)^m + (\alpha - \beta)^{m+1} P(A_{m+1}). \quad (3.15)$$

Eq.(3.15) is the beginning of the "regress of probability-values" that Lewis is talking about. His argument is that, if this series is continued *ad infinitum*, $P(q)$ will always tend to zero, notwithstanding the fact that Reichenbach's correction has been taken into account. This is presumably why Lewis comments: "I disbelieve that it [the addition of the second term] will save his point." Let us see whether Lewis's disbelief is justified.

There are two things that should be noted about (3.15). The first is that it contains only one factor of which the value is unknown. This is $P(A_{m+1})$, i.e. the probability of the first proposition, A_{m+1}, in this finite series. Since all the probabilities in the series are ultimately computed on the basis of this unconditional probability, it seems that we must know its value in order to be able to calculate $P(q)$. The second thing is that, as m gets bigger and bigger, so that the justificatory chain becomes longer and longer, $(\alpha - \beta)^{m+1}$ gets smaller and smaller without limit, finally converging to zero. But of course, if $(\alpha - \beta)^{m+1}$ converges to zero, then $(\alpha - \beta)^{m+1} P(A_{n+1})$ dwindles away to nothing too, for $P(A_{m+1})$ cannot be greater than 1. The right-hand side of Eq.(3.15) is a sum, and if a term in a sum goes to zero, it does not contribute in the limit. With an infinite number of steps, the terms that remain are

$$\begin{aligned} P(q) &= \beta + \beta(\alpha - \beta) + \beta(\alpha - \beta)^2 + \ldots \\ &= \beta \left[1 + (\alpha - \beta) + (\alpha - \beta)^2 + \ldots \right] \\ &= \beta \sum_{n=0}^{\infty} (\alpha - \beta)^n . \end{aligned} \qquad (3.16)$$

Since $\alpha - \beta$ is less than one, the sum here is a convergent geometric series which we can evaluate:

$$P(q) = \frac{\beta}{1 - \alpha + \beta} . \qquad (3.17)$$

In general, (3.17) does not yield zero. For example, if α is $3/4$ and β is $3/8$, then $P(q)$ is $3/5$.[27]

We conclude that Lewis is mistaken. It is not the case that a "regress of probability values" always yields zero. We have just seen an example of such a series, consisting in a sum with an infinite number of terms, that yields a number other than zero. Since Lewis's statement is invalid, it cannot support his main claim that probability statements only make sense if they presuppose certainties.[28]

3.5 A Nonuniform Probabilistic Regress

The counterexample in the previous section is a very special case. For in demonstrating that a probabilistic regress makes sense, we have assumed

---

[27] Eq.(3.17) gives in fact the fixed point of a Markov process. The stochastic matrix governing the process is regular, and the iteration is guaranteed by Markov theory to converge to the solution of the fixed point, $p_* = \beta + (\alpha - \beta)p_*$. However, this quick route to (3.17) only works when the conditional probabilities are the same from step to step: in the general case that we consider in the next section Markov theory does not help, which is why we have not used it here. We shall discuss fixed points more fully in Sections 8.4 and Appendix D.

[28] This example shows that James Van Cleve's defence of Lewis, and thereby his attack on Reichenbach, is mistaken (Van Cleve 1977). Van Cleve argues that an infinite iteration of the rule of total probability must be vicious, because "we must complete it before we can determine any probability at all" (ibid., 328). But our counterexample to Lewis demonstrates that an infinite iteration may well be completable, in the sense that it is convergent and can be summed explicitly, yielding a definite value for $P(q)$.

3.5 A Nonuniform Probabilistic Regress

that the conditional probabilities are uniform, i.e. that they remain the same throughout the entire justificatory chain. Such an assumption is of course rarely fulfilled. It is very uncommon that the degree to which proposition q is probabilistically supported by A_1 is the same as the degree to which A_1 is probabilistically supported by A_2, and so on.

However, it is possible to construct counterexamples without making the assumption that the conditional probabilities are uniform. The rule of total probability relating A_n to A_{n+1} is

$$P(A_n) = P(A_n|A_{n+1})P(A_{n+1}) + P(A_n|\neg A_{n+1})P(\neg A_{n+1}),$$

or, with the abbreviation of the conditional probabilities as α and β, as in the previous section:

$$P(A_n) = \alpha P(A_{n+1}) + \beta P(\neg A_{n+1}).$$

In the nonuniform case the conditional probabilities differ from one link to another, so we have to add an index n to α and β:

$$\begin{aligned}P(A_n) &= \alpha_n P(A_{n+1}) + \beta_n P(\neg A_{n+1}) \\ &= \beta_n + \gamma_n P(A_{n+1}),\end{aligned} \qquad (3.18)$$

where α_n, β_n and γ_n are defined as follows:

$$\begin{aligned}\alpha_n &= P(A_n|A_{n+1}) \\ \beta_n &= P(A_n|\neg A_{n+1}) \\ \gamma_n &= \alpha_n - \beta_n.\end{aligned} \qquad (3.19)$$

Imagine a finite probabilistic chain $A_0, A_1, \ldots, A_{m+1}$, where again A_0 is probabilistically supported by A_1, which is probabilistically supported by A_2, and so on. For notational convenience we have temporarily used A_0 for the target proposition q and A_{m+1} for the grounding proposition p. It is possible to concatenate all the instances of the rule of total probability to yield, for any $m \geq 0$,

$$P(A_0) = \beta_0 + \gamma_0 \beta_1 + \gamma_0 \gamma_1 \beta_2 + \ldots + \gamma_0 \gamma_1 \ldots \gamma_{m-1} \beta_m + \gamma_0 \gamma_1 \ldots \gamma_m P(A_{m+1}). \qquad (3.20)$$

Formula (3.20), of which a proof is given in Appendix A.1, is the nonuniform counterpart of formula (3.15) in the uniform case.

We have seen that, notwithstanding Lewis's opinion, the extension of the finite (3.15) to an infinite chain can be envisaged: in the uniform case the infinite extension is well-defined if the extreme values 0 and 1 for the conditional probabilities are excluded. Does it make sense to extend (3.20) to an infinite number of links? Can a probabilistic regress in the nonuniform case also be well-defined and moreover yield a nonzero value for the target? Again, one example is enough to refute Lewis's argument in this more general setting, and here it is:

$$\alpha_n = 1 - \frac{1}{n+2} + \frac{1}{n+3}; \quad \beta_n = \frac{1}{n+3}; \quad \gamma_n = 1 - \frac{1}{n+2}. \quad (3.21)$$

In (3.21) α_n and β_n depend nontrivially on n. The resulting infinite series is not a geometric series, as it was in the uniform case that was introduced in Section 3.4. Nevertheless, as is shown in Appendix A.5, when we insert the formulae (3.21) into (3.20) we can work out the sum explicitly, obtaining

$$P(A_0) = \tfrac{3}{4} - \tfrac{2m+5}{2(m+2)(m+3)} + \tfrac{1}{m+2} P(A_{m+1}). \quad (3.22)$$

In the limit that m goes to infinity, the second and the third terms on the right-hand side of (3.22), namely $\frac{2m+5}{2(m+2)(m+3)}$ and $\frac{1}{m+2} P(A_{m+1})$, both go to zero. Thus only the term $\tfrac{3}{4}$ survives in the limit, so that $P(A_0)$, that is the probability of the target, $P(q)$, equals $\tfrac{3}{4}$. Here then is a new and more general case that invalidates Lewis's argument that an infinite probabilistic regress must yield zero.

3.6 Usual and Exceptional Classes

The above examples not only illustrate that Lewis was mistaken, but also that a probabilistic regress can have a limit and in that sense be benign. But what are the conditions under which this is so? When exactly does a probabilistic regress yield a well-defined value for the target proposition?

In general there exist two conditions. Each of them is necessary, and together they are sufficient. Look again at our finite nonuniform chain, (3.20):

$$P(A_0) = \beta_0 + \gamma_0 \beta_1 + \gamma_0 \gamma_1 \beta_2 + \ldots + \gamma_0 \gamma_1 \ldots \gamma_{m-1} \beta_m + \gamma_0 \gamma_1 \ldots \gamma_m P(A_{m+1}).$$

The right-hand side of this equation consists of two parts, namely the sum of conditional probabilities,

3.6 Usual and Exceptional Classes

$$\beta_0 + \gamma_0\beta_1 + \gamma_0\gamma_1\beta_2 + \ldots + \gamma_0\gamma_1\ldots\gamma_{m-1}\beta_m,$$

and the remainder term,

$$\gamma_0\gamma_1\ldots\gamma_m P(A_{m+1}).$$

The first condition for a benign probabilistic regress is that the series of conditional probabilities converges in the limit. The second condition is that, as m is taken to infinity, the remainder term goes to zero.

As we prove in Appendix A.3, the first condition is always satisfied, given that we assume probabilistic support, i.e. the constraint $P(A_n|A_{n+1}) > P(A_n|\neg A_{n+1})$ for all n. No matter whether we are dealing with uniform or with nonuniform conditional probabilities, the infinite series

$$\beta_0 + \gamma_0\beta_1 + \gamma_0\gamma_1\beta_2 + \gamma_0\gamma_1\gamma_2\beta_3 + \ldots, \quad (3.23)$$

always converges. However, the matter is different as far as the second condition is concerned. This condition is satisfied in the uniform situation (with the restriction that α is not equal to one and β is not equal to zero), but it is not always satisfied in the nonuniform situation. We shall call the class of cases where both conditions are fulfilled *the usual class*.[29] In the usual class the probability of the target is equal to the following convergent series of terms, each of which is a function of the conditional probabilities only:

$$P(q) = \beta_0 + \gamma_0\beta_1 + \gamma_0\gamma_1\beta_2 + \gamma_0\gamma_1\gamma_2\beta_3 + \ldots. \quad (3.24)$$

The class of cases in which only the first requirement is fulfilled we will call *the exceptional class*. Regresses in the exceptional class do not furnish counterexamples to Lewis's conclusion; but those in the usual class, on the other hand, do so, on condition that at least one of the β_n is nonzero.

When does a nonuniform probabilistic regress fall within the exceptional class? For our purpose this question is of course important, since it creates the watershed between probabilistic regresses which are benign (in the sense that they yield an exact and well-defined value for the target) and those that are not (in the sense that they only yield such a number if they have a first

---

[29] In the usual class the infinite series (3.23) converges even if one relaxes the condition of probabilistic support. However, since we are interested in justification, of which probabilistic support is a necessary condition, this extension of the domain of convergence is not required for our purposes. Moreover, the condition of probabilistic support is needed for our conception of epistemic justification as a trade-off (see Chapter 5) as well as for convergence in the probabilistic networks discussed in Chapter 8.

member, a ground). Clearly the answer to this question depends on whether the remainder term vanishes in the limit. We have seen that this will be the case if the factor $\gamma_0 \gamma_1 \ldots \gamma_m$ vanishes as m goes to infinity. For then the remainder term $\gamma_0 \gamma_1 \ldots \gamma_m P(A_{m+1})$ will die out, since $P(A_{m+1})$, the probability of the grounding proposition, cannot be greater than one.

But when exactly does $\gamma_0 \gamma_1 \ldots \gamma_m$ go to zero? That is the key question. As we show in Appendix A.4, the answer depends entirely on the asymptotic behaviours of α_n and β_n. The factor $\gamma_0 \gamma_1 \ldots \gamma_m$ goes to zero if and only if α_n does not tend to one more quickly than $1/n$ tends to zero, *or* if β_n does not tend to zero more quickly than $1/n$ tends to zero. If at least one of these disjuncts applies, then the nonuniform probabilistic regress falls within the usual class. It then yields a unique probability value for the target proposition, A_0 or q, which does not depend on an inaccessible unconditional probability at infinity. That is, it does not depend on the value of $P(A_{m+1})$ — or $P(p)$ — in Eq.(3.20) in the limit that m goes to infinity.[30] A nonuniform probabilistic regress within this usual class constitutes a counterexample to Lewis's argument. A specific instance is provided by the example (3.21), for this lies in the usual class, since the remainder term in (3.22), $\frac{1}{m+2}P(A_{m+1})$, goes to zero as m goes to infinity. In this limit the right-hand side of (3.22) tends to $\frac{3}{4}$.

If, however, α_n goes to one very quickly *and* β_n goes to zero very quickly as n tends to infinity, more quickly in fact than $1/n$ tends to zero, then the nonuniform probabilistic regress belongs to the exceptional class. In this case the regress does not result in a unique, well-defined probability value for the target proposition, since the unknown probability of the ground still plays a significant role. The regress is now vicious in the sense that the probability of the target depends in part on the inaccessible ground, and it would not form a counterexample to Lewis's foundationalist argument.

An example of a regress in the exceptional class is as follows,

$$\beta_n = \frac{1}{(n+2)(n+3)} \qquad \gamma_n = 1 - \frac{1}{(n+2)^2}, \qquad (3.25)$$

so that

$$\alpha_n = \beta_n + \gamma_n = 1 - \frac{1}{(n+2)^2(n+3)}.$$

Here $1 - \alpha_n$ and β_n both tend to zero as n tends to infinity more quickly than $\frac{1}{n}$ tends to zero, which shows that the example is indeed a member of

---

[30] That the resulting system is consistent, in the sense that there exists at least one assignment of probabilities for all possible conjunctions of the propositions A_n, has been demonstrated by Frederik Herzberg (Herzberg 2013).

3.6 Usual and Exceptional Classes

the exceptional class. In Appendix A.6 we work out the expression for the probability of the target proposition, obtaining

$$P(A_0) = \tfrac{3}{8} - \tfrac{2m+5}{4(m+2)(m+3)} + \tfrac{1}{2}\tfrac{m+3}{m+2}P(A_{m+1}). \tag{3.26}$$

In this case the remainder term, $\tfrac{1}{2}\tfrac{m+3}{m+2}P(A_{m+1})$, does not vanish in the limit. It becomes formally one half times the limit of $P(A_{m+1})$ as m tends to infinity, which is ill-defined.

A probabilistic regress in the exceptional class is characterized by the fact that it is actually very close to a regress of entailments, i.e. to the 'classical' regress, in which A_{n+1} entails A_n for all n. It is therefore to be expected that a straightforward classical regress will also fail to provide us with a counterexample to Lewis's claim, and this is indeed the case. Here is how a classical regress looks in our probabilistic formalism. If A_{n+1} entails A_n for all n, then

$$\alpha_n = P(A_n|A_{n+1}) = 1;$$

and it is shown in Appendix A.7 that (3.20) reduces in this case to

$$P(\neg A_0) = \gamma_0 \gamma_1 \ldots \gamma_m P(\neg A_{m+1}), \tag{3.27}$$

for any m. We have to consider various possibilities for the behaviour of

$$\beta_n = P(A_n|\neg A_{n+1})$$

as n tends to infinity. If β_n were to tend to zero no more quickly than $1/n$ does, the product $\gamma_0 \gamma_1 \ldots \gamma_m$ in (3.27) would tend to zero as m tends to infinity, so $P(\neg A_0) = 0$, irrespective of the behaviour of $P(\neg A_{m+1})$. Moreover it follows also that $P(\neg A_n) = 0$ for all n, which means that β_n is not defined. This is inconsistent, so we conclude that after all β_n must tend to zero more quickly than $1/n$. But then the product $\gamma_0 \gamma_1 \ldots \gamma_m$ tends to some non-zero limit, and so $P(\neg A_0)$ is not uniquely determined, since $P(\neg A_{m+1})$ can be assigned no particular limit as m goes to infinity. The regress of entailments, or implications, is thus necessarily in the exceptional class.

A very special case is when

$$\beta_n = P(A_n|\neg A_{n+1}) = 0 \tag{3.28}$$

for all n. We have then $P(\neg A_0) = P(\neg A_n)$ for all n, so all the probabilities, $P(A_n)$, have the same, undetermined value. Eq.(3.28) implies that $P(\neg A_n|\neg A_{n+1}) = 1$, which is to say that $\neg A_{n+1}$ entails $\neg A_n$, which of course means that A_n entails A_{n+1} (up to measure zero). If $\alpha_n = 1$ and $\beta_n = 0$, then

A_n implies, and is implied by A_{n+1}: there is a regress of bi-implication all the way along the chain. All the probabilities are the same, but the value is undetermined by the regress. Such a regress of bi-implication is vicious in our sense, for here the truth value of the target cannot be determined in the absence of the truth value of the first member.

To summarize, the system of conditional probabilities belongs to the usual class if and only if $1 - \alpha_n$ *or* β_n do not tend to zero more quickly than $1/n$ tends to zero. On the other hand, if $1 - \alpha_n$ *and* β_n both tend to zero more quickly than $1/n$, then the system belongs to the exceptional class, and the unconditional probabilities of the propositions are not determined. The situation in which α_n is nearly one, and β_n is nearly zero, is close to the case of bi-implication. We therefore might call the exceptional class the case of *quasi-bi-implication*.

3.7 Barbara Bacterium

In this chapter we have introduced the concept of a probabilistic regress, that is an epistemic chain of the form

$$q \longleftarrow A_1 \longleftarrow A_2 \longleftarrow A_3 \longleftarrow A_4 \ldots$$

where the arrow is interpreted in terms of probabilistic support. We examined Lewis's view that such a regress is absurd, since it allegedly implies that the probability of q is zero. According to Lewis, the only way to avoid the absurdity was to stop at a proposition, p, which is certain:

$$q \longleftarrow A_1 \longleftarrow A_2 \longleftarrow A_3 \longleftarrow A_4 \ldots \longleftarrow p.$$

We have opposed Lewis's argument by giving counterexamples, i.e. probabilistic regresses which yield a unique, nonzero probability value for the target. Some of these regresses were based on uniform conditional probabilities, others on nonuniform ones.

All our counterexamples were abstract. This is somewhat unfortunate, since a familiar objection to infinite regresses is that they are not concrete and lack practical relevance. The objection becomes even more pressing if one distinguishes (as we did not do here but will do in later chapters) between propositions and beliefs. Propositions are abstract entities, but beliefs are propositional attitudes that people really have. Whereas the idea of an infinite propositional regress might sound not unreasonable, an infinite dox-

3.7 Barbara Bacterium

astic regress seems a contradiction in terms. Where could we ever find a doxastic series of infinite length?

In the next chapters we will come back to this objection, and then we will also discuss the distinction between a propositional and a doxastic regress. At this juncture we will restrict ourselves to showing that a probabilistic regress of propositions also is relevant to a real-life situation.

Imagine that we are trying to develop a new medicine to cure a disease. In this connection, we want to know whether a particular bacterium has a certain trait, T. Bacteria reproduce asexually, so one parent, the 'mother' bacterium, alone produces offspring. After having carried out many experiments, one day we take from a batch a particular bacterium, which we call Barbara. From our experiments we know that the probability that Barbara has T is considerably greater if her mother has T than if her mother lacks it. So if q is 'Barbara has T' and A_1 is 'Barbara's mother has T', then we can say that A_1 probabilistically supports q. It is not certain that Barbara has T if her mother has the trait, but on the other hand Barbara could have T even if her mother does not have it. Thus $1 > P(q|A_1) > P(q|\neg A_1) > 0$.

The unconditional probability of Barbara having T is given by

$$P(q) = P(q|A_1)P(A_1) + P(q|\neg A_1)P(\neg A_1).$$

Whereas the conditional probabilities in this equation, $P(q|A_1)$ and $P(q|\neg A_1)$, may be assumed to have been determined from our experiments, obtaining $P(A_1)$ is a problem. What is the probability that Barbara's mother has T? We know that it is given by

$$P(A_1) = P(A_1|A_2)P(A_2) + P(A_1|\neg A_2)P(\neg A_2),$$

where $P(A_2)$ is the probability that Barbara's grandmother has T, which in turn is conditioned by $P(A_3)$, the probability that Barbara's great-grandmother has T.[31]

It will be clear that we can only compute $P(q)$ if we know $P(A_3)$. And the situation remains the same, even if we add more and more instances of the rule of total probability, going further and further back in Barbara's ancestry. It seems we are only able to compute the probability that Barbara has T if we know what is the unconditional probability that her primordial mother had T. So at first sight it looks as though foundationalists are right: if q is probabilistically justified by A_1, which is probabilistically justified by A_2,

---

[31] In the reading of Pastin the probability intended by Lewis would be $P(q \wedge A_1)$, see footnote 20. But this is neither the probability of interest nor does it fit what is at stake in the debate between Lewis and Reichenbach.

et cetera, then we have to know for sure the probability of the grounding proposition in order to be able to calculate the probability of q.

This impression, intuitive as it may seem, is however incorrect, and we have already seen why. The chain $q \longleftarrow A_1 \longleftarrow A_2 \longleftarrow A_3$ leads to:

$$P(q) = \beta + \beta(\alpha - \beta) + \beta(\alpha - \beta)^2 + (\alpha - \beta)^3 P(A_3),$$

see (3.14). Going infinitely far back into Barbara's ancestry, we obtain (3.16):

$$P(q) = \beta + \beta(\alpha - \beta) + \beta(\alpha - \beta)^2 + \ldots.$$

This does not have a grounding proposition p. A primordial mother of Barbara makes no contribution, yet we are able to calculate the probability that Barbara herself has T, and this probability, notwithstanding Lewis's opinion, is not zero.

Let A_n be the proposition: 'Barbara's ancestor in generation n has T'. Let the probability that a bacterium has T if her mother has T be 0.99, and the probability that a bacterium has T if her mother lacks it be 0.02. So $\alpha = P(A_n|A_{n+1}) = 0.99$, $\beta = P(A_n|\neg A_{n+1}) = 0.02$, and hence $\gamma = \alpha - \beta = 0.97$. Now (3.16) becomes:

$$P(q) = \frac{\beta}{1-\gamma} = \frac{\beta}{1-\alpha+\beta},$$

in agreement with (3.17). With the numbers chosen for α and β, we can now calculate the probability that Barbara has T: it is $\frac{2}{3}$.

The foregoing example made use of uniform conditional probabilities. As an example of a nonuniform probabilistic regress, suppose that an effect of the increasing pollution of the nutrient, as a result of the growing mass of bacteria in it, is that the probability of a bacterium having T increases as time goes on, quite independently of whether the mother bacterium has T. For example, if $\alpha_n = P(A_n|A_{n+1}) = a + b^{n+1}$ and $\beta_n = P(A_n|\neg A_{n+1}) = b^{n+1}$, where a and b are positive numbers such that $a + b < 1$, then α_n and β_n are different from generation to generation, although $\gamma_n = a$ is constant. Note that, since b is less than one, the factor b^{n+1} increases as n decreases, so in Barbara's remote ancestry there was little pollution, but it increases from generation to generation until Barbara herself appears on the scene. Eq.(3.20) once more reduces to a finite geometric series that can be summed:

$$P(q) = b\left[1 + ab + (ab)^2 + \ldots (ab)^m\right] + a^{m+1}P(A_{m+1})$$
$$= b\frac{1-(ab)^{m+1}}{1-ab} + a^{m+1}P(A_{m+1}).$$

3.7 Barbara Bacterium

In the case of an infinite number of generations, since $(ab)^{m+1}$ and a^{m+1} both vanish in the limit of infinite m, we find

$$P(q) = \frac{b}{1-ab}. \qquad (3.29)$$

For example, if $a = \frac{1}{3}$ and $b = \frac{3}{5}$, we find from (3.29) that $P(q) = \frac{3}{4}$.

One might object that our argument so far is still not very realistic, to put it mildly. For a start, the assumption that conditional probabilities are known as precise numbers is a travesty of what is attainable in scientific practice. In real experiments the conditional probabilities are imprecise, merely being known to lie within some specified interval, and as a result, the unconditional probability of the target, too, is subject to measurement error.

Fortunately, when the conditional probabilities are uniform, as for example in the case of Barbara, then it is relatively easy to determine the interval within which the target probability must lie. For suppose that $P(A_n|A_{n+1})$ is in the interval $[\alpha_m, \alpha_M]$, and $P(A_n|\neg A_{n+1})$ is in the interval $[\beta_m, \beta_M]$. It can be shown that expression (3.17) for $P(q)$ is an increasing function of both α and of β;[32] and this means that the uncertainty in $P(q)$ is given by

$$\frac{\beta_m}{1 - \alpha_m + \beta_m} < P(q) < \frac{\beta_M}{1 - \alpha_M + \beta_M},$$

on condition that $\alpha_M - \beta_m < 1$.

In the more general case where the conditional probabilities are not uniform, the calculation of the uncertainty in the value of $P(q)$ is a little more intricate. However, since the condition of probabilistic support is in force, all the terms in Eq.(3.23) are positive, and it can be done without too much effort. One has to minimize and maximize each term, within the experimental error bounds, in order to obtain lower and upper bounds on $P(q)$.

Even so, one might still feel the urge to protest that we are not dealing with real life situations. No bacterium has an infinite number of ancestor bacteria, if only because of the fact of evolution from more primitive algal slime,

---

[32] The partial derivatives of $\frac{\beta}{1-\alpha+\beta}$ with respect to α and β are both positive:

$$\frac{\partial}{\partial \alpha} \frac{\beta}{1-\alpha+\beta} = \frac{\beta}{(1-\alpha+\beta)^2} > 0$$

$$\frac{\partial}{\partial \beta} \frac{\beta}{1-\alpha+\beta} = \frac{1-\alpha}{(1-\alpha+\beta)^2} > 0.$$

which had grown out of earlier life forms, which sprang from inanimate matter, which originated in a supernova explosion, and so on.

This is of course true, and it makes short shrift of any remaining thought about a beginning in the form of a first bacterium.[33] For our approach, however, the issue is moot. The reason is that the further away a node in the chain is from the target, the smaller its influence on the target becomes. Applied to Barbara: long before we reach the stage where her ancestor bacteria evolve from more primeval life forms, they have become totally irrelevant to the question whether Barbara has T. This phenomenon we call 'fading foundations', and it is explained in the next chapter.

---

[33] Sanford 1975, 1984; Rescher 2010, 56.

Open Access This chapter is licensed under the terms of the Creative Commons Attribution 4.0 International License (http://creativecommons.org/licenses/by/4.0/), which permits use, sharing, adaptation, distribution and reproduction in any medium or format, as long as you give appropriate credit to the original author(s) and the source, provide a link to the Creative Commons license and indicate if changes were made.

The images or other third party material in this chapter are included in the chapter's Creative Commons license, unless indicated otherwise in a credit line to the material. If material is not included in the chapter's Creative Commons license and your intended use is not permitted by statutory regulation or exceeds the permitted use, you will need to obtain permission directly from the copyright holder.

Chapter 4
Fading Foundations and the Emergence of Justification

Abstract
A probabilistic regress, if benign, is characterized by the feature of fading foundations: the effect of the foundational term in a finite chain diminishes as the chain becomes longer, and completely dies away in the limit. This feature implies that in an infinite chain the justification of the target arises exclusively from the joint intermediate links; a foundation or ground is not needed. The phenomenon of fading foundations sheds light on the difference between propositional and doxastic justification, and it helps us settle the question whether justification is transmitted from one link in the chain to another, as foundationalists claim, or whether it emerges from a chain or network as a whole, as is maintained by coherentists and infinitists.

4.1 Fading Foundations

In the previous chapter we have introduced the idea of a probabilistic regress, and we have seen that such regresses are in general unproblematic: they mostly have a calculable limit, thus providing the target proposition, q, with a unique probability value. In all but a few exceptional cases there is no conceptual problem in saying that q is probabilistically supported by an epistemic chain of infinite length.

An important part of our argument concerned the rôle of the foundational or grounding proposition, p. In calculating the unconditional probability of the target, q, we managed to eliminate all the unconditional probabilities — except that of p. The factor $P(p)$ remained the only term in the chain of which the value was unknown. Consider the finite chain

4 Fading Foundations and the Emergence of Justification

$$q \longleftarrow A_1 \longleftarrow A_2 \longleftarrow \ldots \longleftarrow A_{m-1} \longleftarrow A_m \longleftarrow p,$$

where q is probabilistically supported by A_1, which is probabilistically supported by $A_2, \ldots$, and so on, until A_m, which is probabilistically supported by the grounding proposition or belief p.

In any finite chain, we need to know the value of value of $P(p)$ in order to calculate $P(q)$. However, the *importance* of the unknown $P(p)$ for the probability of the target, $P(q)$, lessens as m gets bigger. If the chain is very short, consisting only of two propositions, q and p, then the importance of $P(p)$ for $P(q)$ is at its height: all the support for q comes from p (together with the pair of conditional probabilities that connect the one to the other). But now imagine that the chain is a little bit longer, consisting of three propositions:

$$q \longleftarrow A_1 \longleftarrow p.$$

In terms of nested rules of total probability this becomes:

$$P(q) = P(q|\neg A_1) + [P(q|A_1) - P(q|\neg A_1)]\{P(A_1|\neg p) \\ + [P(A_1|p) - P(A_1|\neg p)]P(p)\}. \quad (4.1)$$

In (4.1) the importance of $P(p)$ has somewhat decreased. It is still the case that it largely determines $P(q)$, but the influence of the conditional probabilities has become greater. In general it is so that, as the chain becomes longer, the support provided by the totality of the conditional probabilities increases, while that given by the foundation decreases. In other words, as m in A_m grows larger and larger, a law of diminishing returns come into force: the influence of $P(p)$ on $P(q)$ tapers off with each link, until it finally fades away completely. In the limit that m tends to infinity, all the probabilistic support for q comes from the conditional probabilities together, and none from the ground or foundation. This characteristic, that is essential to a probabilistic regress as we defined it, we call the feature of *fading foundations*. As we add more and more links to the chain the influence of $P(p)$ tails off, and $P(q)$ draws closer and closer to its final value.

The feature of fading foundations can be illustrated by our story about Barbara bacterium in the previous chapter. Recall that q is the proposition 'Barbara has trait T', A_n is 'Barbara's ancestor in the nth generation has T', and p is 'Barbara's primordial mother has T'. Now imagine that long and extensive empirical research in our laboratory has taught us that the probability that a bacterium has T is 0.99 when her mother has T, and that it is 0.04 when her mother lacks T:

4.1 Fading Foundations

$$P(q|A_1) = P(A_1|A_2) = \ldots = P(A_{m-1}|A_m) = P(A_m|p) = 0.99$$
$$P(q|\neg A_1) = P(A_1|\neg A_2) \ldots = P(A_{m-1}|\neg A_m) = P(A_m|\neg p) = 0.04$$

Let us further take for the unconditional probability of p the value 0.7. With the numbers we have chosen for the conditional probabilities, 0.99 and 0.04, the computed values for the unconditional probability of q are listed in the following table:

Table 4.1 Probability of q when the probability of p is 0.7

Number of A_n	1	2	5	10	25	50	75	100	∞
Probability of q	.710	.714	.726	.743	.774	.793	.798	.799	.8

The first entry in this table refers to the chain $q \longleftarrow A_1 \longleftarrow p$, where there is only one A. With the values that we have chosen in our example, the probability of the target proposition q yielded by this chain is 0.709. The second entry corresponds to the chain $q \longleftarrow A_1 \longleftarrow A_2 \longleftarrow p$. Here there are two A's, so the probabilistic support for q has grown, resulting in a probability for q that is somewhat higher, namely 0.714. The third entry refers to a chain of seven propositions: the target proposition q, five A's and the grounding proposition p. The support is still further augmented, and the probability of q equals 0.726. By including more and more A's we observe that the probabilistic support for q grows. The final entry corresponds to the situation where the chain is infinitely long. Here the probabilistic support for q has reached its maximum, culminating in the unconditional probability $P(q) = 0.8$. The latter can considered to be the 'real' value for the probability of q relative to the numbers chosen for the conditional probabilities.[1]

But now look at the second table, 4.2, where the conditional probabilities are the same as in Table 4.1, but where the unconditional probability of p is 0.95. There are two things that should be noted about these two tables. Firstly, the probability of q in Table 4.2 culminates in a limiting value that is the same as that in Table 4.1, namely 0.8. Secondly, while the numbers in

---

[1] In this table as well as in the following one, the values of the conditional probabilities are uniform, remaining the same throughout the chain. As has been explained in the previous chapter, and more in detail in the appendices, this is however not essential to the phenomenon of fading foundations. The argument goes through, in the usual class, when the values of the conditional probabilities differ from link to link.

Table 4.1 steadily increase as the number of links becomes larger, those in Table 4.2 go down. How can we understand these facts?

Table 4.2 Probability of q when the probability of p is 0.95

Number of A_n	1	2	5	10	25	50	75	100	∞
Probability of q	.935	.929	.910	.885	.840	.811	.803	.801	.8

The answer is provided by the feature of fading foundations. As the chain lengthens, the role of the foundation p becomes less and less important until it dies out completely. At the end of the day, the probability of q is fully determined by the conditional probabilities; everything comes from them and the influence of the foundation p has completely disappeared from the picture. The reason why the numbers in Table 4.1 go up, while those in Table 4.2 go down, is because in the first case the probability p is lower than the final real value of $P(q)$, relative to the chosen conditional probabulities, while in the second case it is higher. This is exactly what is to be expected as the foundational influence gradually peters out.

Lewis and Russell were right that, in a probabilistic regress, *something* goes to zero if m goes to infinity. However, this 'something' is not the value of $P(q)$, as they thought. Rather it is the influence that the foundation p has on the target q. This is not to say that p itself has become highly improbable, for p may have any probability value at all. It is rather that, in the limit, the effect of the would-be foundation p has faded away completely: the support it gives to q is nil.[2]

4.2 Propositions versus Beliefs

Up to this point we have not distinguished between propositional and doxastic justification: q, the A's, and p could be either propositions or beliefs.

---

[2] The fading influence of the foundation p should not be confused with the familiar washing out of the prior in Bayesian reasoning. In Bayesian updating, the prior probability becomes less and less important under the influence of new pieces of information coming in, until it washes out completely. Although this looks rather like the phenomenon of fading foundations, where the influence of p similarly diminishes, the two phenomena are actually quite different, as we explain in Appendix C.

4.2 Propositions versus Beliefs

However, it has often been pointed out that the distinction *is* relevant when we talk about justification, especially if we discuss the possibility of infinite justificatory chains. In this section we will look at a debate between Michael Bergmann and Peter Klein in order to explain how the phenomenon of fading foundations can shed light on the subject.[3]

Bergmann has criticzed Peter Klein's infinitism by arguing that, although propositional justification might go on and on, doxastic justification must always come to a stop; infinite epistemic chains and doxastic justification simply seem incompatible.[4] In a reply to Bergmann, Klein has acknowledged that, unlike propositional justification, doxastic justification is always finite. As he wryly notes, "We get tired. We have to eat. We have satisfied the enquirers. We die".[5] He does not regard this as a difficulty for infinitism, however, since the stop is merely contextual or pragmatic. According to Klein, "doxastic justification is parasitic on propositional justification": in principle it can go on, but in practice it ends.[6]

Bergmann, however, believes that Klein's position is untenable, arguing as follows.[7] In order to reject foundationalism, Klein must endorse the following view:

K_1: For a belief B_i to be doxastically justified, it must be based on some other belief B_j.

Bergmann then introduces

---

[3] See Peijnenburg and Atkinson 2014b. We will say a bit more about the distinction between propositional and doxastic justification in the next chapter, when we discuss Klein's reply to the notorious finite mind objection. For the difference between propositional and doxastic justification, see also Turri 2010.

[4] Bergmann 2007. Jonathan Kvanvig has argued that Klein's infinitism has difficulties not only accounting for doxastic justification, but for propositional justification too (Kvanvig 2014). We will briefly come back to Kvanvig's criticism in the next chapter.

[5] Klein 2007a, 16. See Poston 2012, which contains a proposal for emerging justification on the basis of Jonathan Kvanvig's INUS conditions.

[6] Ibid., 8. Michael Williams (Williams 2014, 234-235) has noted that the distinction between doxastic and propositional justification was introduced by Roderick Firth (Firth 1978). He recalls that Firth, too, claims that doxastic justification is parasitic on propositional justification, but argues that Firth attaches a completely different meaning to this claim than does Klein. As Williams sees it, Klein tries to combine an infinitist conception of propositional justification with a contextual conception of doxastic justification — a venture that, according to Williams, is doomed to failure (Williams 2014, 236-238).

[7] Bergmann 2007, 22-23.

K_2: A belief B_i can be doxastically justified by being based on some other belief B_j only if B_j is itself doxastically justified.

and subsequently tries to catch Klein on the horns of a dilemma. Klein must either accept or reject K_2. If he rejects it, then he must maintain that a belief B_i can be doxastically justified by another belief B_j even if the latter is itself unjustified. This would turn Klein into a defender of what Bergmann calls *the unjustified foundations view* — an outlook that is not particularly Kleinian, to say the least. On the other hand, if Klein accepts K_2 along with K_1, then he would run the risk of becoming a sceptic. For then "he is committed to requiring for doxastic justification an infinite number of actual beliefs. ... But it seems completely clear that none of us has an infinite number of actual beliefs".[8]

The phenomenon of fading foundations points to an escape route out of this dilemma, for it shows that there is another way to reject K_2. If doxastic justification indeed draws on propositional justification, as Klein claims, then the justification that one belief gives to another also diminishes as the distance between them increases. That is to say, a belief B_1 can be doxastically justified by a chain of other beliefs, B_2, B_3, to B_n, such that:

1. each B_m is conditionally justified by B_{m+1}, where $2 \leq m \leq n-1$;
2. B_n may be justified by another belief, or may justify itself, or may be unjustified;
3. the effect of B_n on B_1 becomes smaller as n becomes bigger and bigger.

In the limit that n goes to infinity, the justificatory support given by B_n to B_1 vanishes completely. In that case it does not matter for the doxastic justification of B_1 whether B_n is justified or not: B_1 can still be doxastically justified. Klein and Bergmann are of course right that we cannot forever go on justifying our beliefs. But the phenomenon of fading foundations manifests itself already in chains of finite length. Often we need only a few links to observe that the influence of the foundational belief on the target belief has diminished considerably. Of course, we can only be sure of what we seem to be observing in a finite chain if there exists a convergence proof for the corresponding infinite series, and a proof that the remainder term goes to zero: there needs to be knowledge of what happens in the infinite case in order for us to be certain that what we see in the finite case is a robust phenomenon rather than a mere fluctuation. But as we have seen such a proof can be provided. Klein, too, argues that "rejecting K_2 does not entail endorsing an

---

[8] Bergmann 2007, 23. See also Bergmann 2014.

4.2 Propositions versus Beliefs

unjustified foundationalist view" (Klein 2007b, 28). His argument is different from ours, in that it refers, among other things, to a reason's availability. We however believe that our reasoning about fading foundations can capture Klein's most important intuitions, and we will come back to availability in the next chapter.

Let us sum up. In doxastic justification the choice is not between indefinitely going on and the unjustified foundations view. There is a third possibility, provided by what we know about infinite chains. Once we have recognized that any justification that B_n gives to B_1 diminishes as the distance between the two is augmented, we might decide to stop at B_n because the justificatory contribution that any further belief would bestow on B_1 is deemed to be too small to be of interest. When exactly a justificatory contribution is considered to be negligible depends on pragmatic considerations, but our two tables show that we are able to make these considerations as precise as we wish.

This third possibility goes unnoticed in the debate between Bergmann and Klein. Because the fact of fading foundations has not been taken into account, they fail to realize that the expression 'stopping at a belief B_n' can have more meanings than those that have been envisioned in the literature. It need not mean 'making an arbitrary move', as some coherentists have claimed. Nor need it imply that B_n is taken to be unjustified or self-justified. Rather, an agent can decide to stop at a belief B_n because she realizes that, for her purposes, B_{n+1} has become irrelevant for the justification of B_1. She finds that the degree of justification conferred upon B_1 by her beliefs B_2 to B_n is accurate enough, and she feels no call to make it more accurate by taking B_{n+1} into account. For her, the justificatory contribution that B_{n+1} gives to B_1 has become negligible, and with our tables she can precisely identify a point at which the role of B_n is small enough to be neglected, where we use the word 'justificatory' as before as meaning probabilistic support plus something else.

In this way we have given a more precise meaning to contextualist considerations that have been often expressed. For example Klein:

> The infinitist will take the belief that q to be doxastically justified for S just in case S has engaged in providing 'enough' reasons along the path of endless reasons. ...How far forward ...S need go seems to me a matter of the pragmatic features of the epistemic context.[9]

---

[9] Klein 2007a, 10.

> We don't have to traverse infinitely many steps on the endless path of reasons. There just must be such a path and we have to traverse as many as contextually required.[10]

And Nicholas Rescher:

> In any given context of deliberation the regress of reasons ultimately runs out into 'perfectly clear' considerations which are (contextually) so plain that there just is no point in going further. ... Enough is enough.[11]

Our method differs however from what Klein and Rescher seem to have in mind. As we will explain in more detail in 5.3, where we argue for a view of justification as a kind of trade-off, the level of accuracy of the target can be decided upon in advance. Whether this level will be reached after we have arrived at proposition number three, four, sixteen, or more, depends on the structure of the series and on the chosen level. In no way does it depend on the question of how obvious proposition number three, four, sixteen, etc. is. Even if the proposition at issue is very obvious, and thus has a high probability, its contribution to the justification of the target might be small enough to be neglected. This is different from the contextualism of Klein and Rescher, according to which an agent stops when the next belief in the chain is sufficiently obvious and itself not in need of justification.

4.3 Emergence of Justification

It has been said that foundationalists and anti-foundationalists (that is coherentists and infinitists) conceive justification differently: the former gravitate towards an atomistic concept of justification, whereas the latter see it as a holistic notion.[12] Consequently, foundationalists regard justification as a property that can be *transmitted* or *transferred* from one proposition to another. The idea here is that justification somehow arises as a quality attached to a particular proposition, notably to the ground p, and then via inference is conveyed to the neighbouring proposition. The inferences themselves in no way affect the property that they transfer. They are just conduits, as McGrew and McGrew would have it, completely neutral in character, like wifi connecting two computers.[13]

---

[10] Ibid., 13.
[11] Rescher 2010, 47.
[12] Sosa 1980; Bonjour 1985; Dancy 1985.
[13] McGrew and McGrew 2008.

4.3 Emergence of Justification

Anti-foundationalists, on the other hand, have a different outlook. For them justification is not a property that is transmitted from one link in the chain to another; rather it *emerges* gradually from the chain as a whole. In the words of Peter Klein:

> Foundationalists think of propositional justification as a property possessed autonomously by some propositions which, by inference, can then be transmitted to another proposition — just as a real property can be transmitted from one owner to another once its initial ownership is established. But of course, the infinitist, like the emergent coherentist, does not paint this picture of propositonal justification. ... [T]he infinitist conceives of propositional justification of a proposition as emerging whenever there is an endless, non-repeating set of propositions available as reasons.[14]

> ... the infinitist does not think of propositional justification as a property that is transferred from one proposition to another by such inference rules. Rather, the infinitist, like the coherentist, takes propositional justification to be what I called an emergent property that arises in sets of propositions.[15]

However, infinitists and coherentists experience great difficulty in explaining emergence. What exactly does it mean to say that justification emerges from a chain of propositions? How precisely does justification gradually arise from a chain or a web of beliefs? Champions of emergence illustrate their views by invoking arresting images, such as Neurath's boat or Sosa's raft. Although such metaphors are striking and helpful, they fail to inform us how exactly emergence can occur. It is one thing to claim that justification can emerge, but quite another to come up with a mechanism which explains how this can happen. Yet the latter is what we need. When emergence is called on to save the day for the anti-foundationalist, an account of the mechanism behind it ought to be specified in detail. Without such an account, emergence is in danger of being not much more than a name, and the appeal to it runs the risk of remaining gratuitous or *ad hoc*.

We believe that our concept of probabilistic support can help us here. For it carries with it the idea of fading foundations, which explains how justification can gradually emerge.[16] Look again at Table 4.1. It reveals the justification as it emerges from an infinite chain of reasons, and as a result we see the justification of q materializing in front of our eyes, as it were. The

---

[14] Klein 2007a, 16.

[15] Klein 2007b, 26.

[16] Frederik Herzberg also argues that our notion of probabilistic support can help explaining emergence (Herzberg 2013).

table enables us to give a precise interpretation of what Klein writes about justification as seen by infinitists (recall that for Klein doxastic justification is parasitic on propositional justification):

> ...the infinitist holds that propositional justification arises in sets of propositions with an infinite and non-repeating structure such that each new member serves as a reason for the preceding one. Consequently, an infinitist would seek to increase the doxastic justification of an initial belief – the belief requiring reasons – by calling forth more and more reasons. The more imbedded the initial belief, the greater its doxastic justification.[17]

Thus for Klein justification increases by lengthening the chain. A similar idea has been expressed by Jeremy Fantl:

> The infinitist [claims] that, for any particular series of reasons, the degree of justification can be increased by adding an adequate reason to the end of that series. Infinitism [claims]: ...the longer your series of adequate reasons for a proposition, the more justified it is for you.[18]

Our analysis can give a more precise meaning to these claims by Klein and Fantl. For it makes it clear that phrases like 'the emergence of justification' or 'the increase of justification' are in fact ambiguous. They can mean that, by adding more and more reasons, the value of the unconditional probability of *q becomes larger and larger*. But they can also mean that, by adding more reasons, the value of the unconditional probability of *q draws closer to its final value* (relative to the numbers chosen). It is the latter meaning that we are talking about here. In Table 4.1 it is the case that, every time we add an extra link to the chain, the probability of q rises until it reaches its maximum value. A rising value is however not essential for justification to emerge. This can be appreciated in Table 4.2, where the conditional probabilities are the same as those in Table 4.1, but where the unconditional probability of p is 0.95.

As in Table 4.1, in Table 4.2 the justification of q emerges as the number of A's gets bigger, for now q is, as Klein would say, more imbedded. However, it is not so that the probability of q rises with each step. As we

---

[17] Klein 2007b, 26.

[18] Fantl 2003, 554. Fantl defends infinitism on the grounds that, of all the theories of justification, it is best equipped to satisfy two requirements: the degree requirement ("a theory of the structure of justification should explain why or show how justification is a matter of degree") and the completeness requirement ("a theory of the structure of justification should explain why or how complete justification makes sense") — ibid., 538. That reasoning itself can generate justification has also been advocated by Mylan Engel (2014) and John Turri (2014).

4.3 Emergence of Justification

add more and more reasons, the probability of q gets closer and closer to its final value, but numerically it goes down, namely from 0.935 to 0.8. Klein's phrase "[t]he more imbedded the initial belief, the greater its doxastic justification" or Fantl's phrase "the longer your series of adequate reasons for a proposition, the more justified it is for you" should therefore be properly interpreted. The phrases are correct under the interpretation: the longer the chain that justifies the target q, the more reliable the justification of q is, for the closer the unconditional probability of q is to its real value. What cannot be meant is: the longer the chain that justifies the target q, the greater the unconditional probability of q. The justification of q can ascend in reliability while the probability of q descends in numerical value. So we should be careful about what we mean when we say that justification emerges: we do not mean that the unconditional probability of the target proposition q necessarily increases numerically, rather we mean that this probability gradually moves towards its limit.

So far we have worked under the assumption that the values of $P(p)$ lay strictly between 0 and 1. Indeed, both Tables 4.1 and 4.2 respect this restriction. However, the assumption is neither necessary for fading foundations nor for the emergence of justification. The two tables below illustrate this point.

Table 4.3 Probability of q when the probability of p is 1

Number of A_n	1	2	5	10	25	50	75	100	∞
Probability of q	.981	.971	.947	.914	.853	.814	.804	.801	.8

Table 4.4 Probability of q when the probability of p is 0

Number of A_n	1	2	5	10	25	50	75	100	∞
Probability of q	.078	.114	.212	.345	.589	.742	.784	.796	.8

These tables are based on the same uniform conditional probabilities that we used before, that is 0.99 and 0.04. However, in Table 4.3 the unconditional probability of p is one and in Table 4.4 it is zero. They are extreme values, and admittedly they yield strange consequences. For example, if $P(p) = 0$, then p can scarcely be called a reason for q. And if $P(p) = 1$, then p cannot provide probabilistic support for any proposition (this is the root of the infamous problem of old evidence). Yet the tables reveal how ineffective the rôle of p is in the long run. For even with a $P(p)$ that is zero, the final probability of q is still 0.8; and justification can emerge when the foundation is non-existent. Notwithstanding the extreme values of $P(p)$, the final probability of q is the same, and moreover the same as it was in Tables 4.1 and 4.2.[19]

In sum, we have argued that, in a probabilistic model of epistemic justification, justification is not something that one proposition or belief receives lock, stock and barrel from another. Rather it gradually emerges from the chain as a whole. As the distance between the source p and the target q *increases*, the influence of the unconditional probability of p on the unconditional probability of q *decreases*; in the limit of an infinite chain, the probability of q reaches its final value, and the only contributions to this value come from the infinite set of conditional probabilities. So when we go probabilistic, a law of diminishing returns goes hand in hand with a law of emerging justification: the more the justification of the final proposition materializes, the less is the influence of the grounding proposition.

4.4 Where Does the Justification Come From?

In a finite probabilistic chain, part of the justification comes from the ground and part comes from the conditional probabilities that connect the ground to the target. If the series is infinite, then *all* of the justification is carried by the conditional probabilities, and none by the ground. One might however still be puzzled as to whence the justification comes. If justification does not have its origin in a foundation, then where does it come from? How can we make sense of there being justification without a ground?

Most people agree that having justification somehow involves making contact with the world; as we said in Chapter 2, to call our beliefs justified means acknowledging that they at least remotely indicate how things actually are. If one takes the view that contact with the world requires a ground, and that a ground is apprehended by a basic belief, and that a basic belief involves an unconditional probability, then it is puzzling indeed how infinite chains can do the job. Such a view would however be unduly restrictive. It assumes that notions like 'applying to the real world', 'outside evidence'

---

[19] If $P(p)$ is zero or one, some of the conditional probabilities are not well-defined according to Kolmogorov's prescription. Alternative approaches to probability theory exist however, in which conditional probabilities are the basic quantities, and we will come back to this in the next section. The important point here is that if $P(p) = 1$ then $P(A_m) = P(A_m|p)$, and $P(A_m|\neg p)$, which does not have a Kolmogorovian definition, is not needed as an ingredient in the regress. Similarly, if $P(p) = 0$ then $P(A_m) = P(A_m|\neg p)$, and $P(A_m|p)$ is not needed in the regress.

4.4 Where Does the Justification Come From?

and 'empirical results' only makes sense within a framework of basic beliefs. This is questionable, since conditional probabilities are just as well equipped to carry the empirical burden.

One might object that conditional probabilities are built up from unconditional ones, and that one can only determine their values on the basis of unconditional probabilities. Such a complaint has in fact been made by Nicholas Rescher:

> There is ... a more direct argument against the thesis that one can never determine categorical probabilities but only conditional ones. This turns on the fact that conditional probabilities are by definition no other than ratios of unconditioned ones $P(q|p) = P(q\&p)/P(p)$. So unless conditional probabilities are somehow given by the Recording Angel they can be only be determined (or estimated) via our determination (or estimation) of categorical probabilities. And then if the latter cannot be assessed, neither can the former.[20]

It is true that, within standard probability theory, conditional and unconditional probabilities can be defined in terms of one another. It is also true that Kolmogorov himself saw the unconditional probabilities as the basic elements. However, three considerations should be taken into account here. First, one is free to make another choice, and many philosophers have done so. Rudolf Carnap, Karl Popper, Alan Hájek — they all plump for conditional probabilities as the more useful basic quantities. In fact taking conditional probabilities as primary has certain advantages: one can cover extreme cases that cannot be handled if unconditional probabilities are regarded as being fundamental.[21] Second, we have not claimed that unconditional probabilities can *only* be estimated via infinite regresses involving conditional probabilities: rather we have shown that they *can* be computed in that way. Third and most important, there is no objection whatever to questioning the conditional probabilities in turn. Up to this point we have considered them as being given, but that is only a pragmatic stance, motivated by expository considerations. It is perfectly possible to unpack the conditional probabilities and consider them as targets that are themselves justified by further probabilistic chains. This possibility will be briefly touched upon in Section 6.4

---

[20] Rescher 2010, 40, footnote 18 (we adapted Rescher's notation to ours).

[21] Carnap 1952; Popper 1959; Hájek 2011. Hájek mentions more philosophers who made this choice: De Finetti 1974/1990; Jeffreys 1939/1961; Johnson 1921; Keynes 1921; Rényi 1970/1998. One can define $P(q|p)$ as $P(q \wedge p)/P(p)$ only if $P(p) \neq 0$. If one adopts this Kolmogorovian definition, one is unable to make sense of $P(q|p)$ when $P(p) = 0$. The approach of the philosophers mentioned above is free from this difficulty.

and further explained in Section 8.5. But for the moment we ignore this refinement.

Two final worries remain. First, how do we know that the conditional probabilities in our chain are 'good' ones, i.e. make contact with the world? What is the difference between our reasonings and those occurring in fiction, in the machinations of a liar, or in the hallucinations of a heroin addict? Or, applied to our example about bacteria, how can we distinguish the regress concerning Barbara and her ancestors from a fairy tale with the same structure in which, instead of the inheritable trait T, there is an inheritable magical power, M, to turn a prince into a frog?

The distinction is not far to seek. It lies in the mundane fact that in the former, but not in the latter, the conditional probabilities arise from observation and experiment. Research on many batches of bacteria have established the relevant conditional probabilities, α and β. These conditional probabilities are typically obtained by repeated experiments: they are measured by counting how many 'successes' there are in a given number of trials, and then by dividing one number by the other (e.g. the number of bacteria that carry a trait, divided by the total number of bacteria in a sample). In the fairy tale, on the other hand, the only 'evidence' that M is inheritable is contained in the story itself — outside the tale there is no evidence at all. When it comes to series of infinite length, conditional probability statements are the sole bearers of the empirical load. Together they work to confer upon the target proposition an unconditional probability that expresses the proposition's degree of justification. It is by virtue of the conditional probabilities that an infinite chain is not just an arbitrary construct that displays mere coherence, but rather can provide real justification, albeit of a probabilistic character.

We realize perfectly well that this answer will not convince the confirmed sceptic, but our opponent after all is a particular kind of foundationalist, not the sceptic. We do not have the temerity to aim at refuting the claim that all our perceptions might be illusory, or at outlawing evil demon scenarios, old and new. We simply assume that there is a real world, and that empirical facts can justify certain propositions, or more generally can sanction the probabilities that certain propositions are true. Here we merely take issue with any foundationalist claim to the effect that only basic beliefs or unconditional probabilities can be candidates for connecting world and thought.

That brings us to the second worry. A foundationalist might not be persuaded by the above considerations, arguing that the erstwhile rôle of the basic belief is now being played by the set of conditional probabilities. Indeed, he might claim that we are worse off, for we seem to have traded one

4.4 Where Does the Justification Come From?

basic belief, viz. the remote starting point of the epistemic chain, for an infinite number of conditional probability statements.

We do not want to get involved in a verbal dispute here: we are not objecting to a type of foundationalism that acknowledges the empirical thrust of conditional probabilities as well as the importance of fading foundations. This should not blind us, however, to the difference between conditional probabilities and the traditional basic beliefs. The former are essentially relational in character: they say what is to be expected if something else is the case. The latter are by contrast categorical: they say that something is the case, or that something can be expected with a certain probability. There is a great difference between averring that 'A_n is true' (or that the probability of A_n is large) on the one hand, and holding that 'if A_{n+1} were true, the probability that A_n is true would be α', or 'if A_{n+1} were false, the probability that A_n is true would be β' on the other hand. Conditional probability talk is discourse about relationals and hypotheticals. Our use of an infinite number of conditional probabilities amounts to the introduction of an infinite number of relational statements. If all these statements satisfy the condition of probabilistic support as defined earlier, they can give rise to something that is no longer relational, but categorical. This categorical statement can in turn become the starting point of a new series of relational statements. And if this new series becomes sufficiently long, the influence of the categorical might die out, as we have seen.

The situation is somewhat comparable to what happens in science or in logic.[22] Scientists typically construct mathematical models on the basis of empirical input, and then employ these models to draw new conclusions about the world. Similarly, logicians make inferences on the basis of premises that contain empirical information, thus producing new conclusions as output. In both cases, the output can in turn become the input for other models and inferences. And in neither case can the machinery work without input: logicians need their premises and scientists need their data. Since every assumption that serves as input can itself be questioned in turn, there is in this sense a foundation behind every foundation. One may interpret that as support for foundationalism ('there is always a foundation!') or as support for anti-foundationalism ('every foundation is a pseudo-foundation!'). Rather than let ourselves be drawn into such a debate, it might be more fruit-

---

[22] Gijsbers 2015; Bewersdorf 2015.

ful to see what actually happens. And what happens is that a foundation becomes less important as it recedes from the target.[23]

4.5 Tour d'horizon

Let us take stock. The epistemological regress problem, as we have introduced it in Chapter 1, led to a discussion of epistemic justification in Chapter 2. The idea that epistemic justification has something to do with 'probabilification' is widespread among contemporary epistemologists: practically all agree that 'A_j justifies A_i' at least implies that A_i is made probable by A_j. Yet, as we have been arguing in Chapters 3 and 4, the far-reaching consequences of this unanimity about the regress problem in epistemology have been insufficiently understood.

A few exotic cases excluded, talk about probability is Kolmogorovian talk. One of the theorems of Kolmogorov's calculus is the rule of total probability, which enables us to determine the unconditional probability of q, namely $P(q)$. If $P(q)$ is made probable by an epistemic chain rather than a single proposition, then the value of $P(q)$ is obtained from an iterated rule of total probability. It has often been thought that such an iteration does not make sense if it continues indefinitely, but, as we have seen in Chapter 3, this is simply a mistake. In all but the exceptional cases $P(q)$ can be given a unique and well-defined value, even if the chain that supports it is infinitely long.

The iteration in question is a complex formula that consists of two parts. The first part is a series involving all the conditional probabilities, the second part is what we have called the remainder term, which contains information

---

[23] The phenomenon of fading foundations is not restricted to probabilistic chains in epistemology; it can be proved (although we will not do that here) that it also applies in modified form to infinite chains of propositions that are ranked in the sense of Spohn (Spohn 2012). Moreover, fading foundations occur in non-epistemic causal chains, as long as 'causality' is interpreted probabilistically. This fact may shed light on various philosophical debates, such as the one on rigid designators, i.e. expressions that denote the same object in every possible world. The objects themselves, at least for Saul Kripke, are identified by following causal chains backwards to the moment of baptism when they received their names. Gareth Evans noted a problem with this view: we can use proper names even if the causal chains are broken (his Madagascar-example in Evans 1973). In Addendum (e) to *Naming and Necessity* Kripke comments that he leaves this problem "for further work" (Kripke 1972/1980, 163); but with a probabilistic conception of causality Evans' problem disappears since the rôle and character of rigid designators change.

4.5 Tour d'horizon

about the probability of the grounding proposition. What in this chapter we have called fading foundations arises if and only if the following two requirements are fulfilled:

1. the series involving the conditional probabilities converges
2. the remainder term goes to zero.

The first requirement is always fulfilled if the condition of probabilistic support has been satisfied for the entire chain; that is, if $P(A_i|A_j) > P(A_i|\neg A_j)$ for all the links. The second requirement is only fulfilled if we are dealing with what we have been calling the usual class, i.e. the class of probabilistic regresses that are benign. Informally, this means that the conditional probabilities must not tend too quickly to those appertaining to an entailment. Formally, it means that they comply with

$$\exists c > 0 \ \& \ \exists N > c: \ \forall n > N, \ 1 - \gamma_n > \frac{c}{n}.$$

Whereas conditional probabilities that obey this constraint belong to the usual class, those that violate it make up the exceptional class. The latter we also call the class of *quasi-bi-implication*. The conditional probabilities in this class resemble bi-implications, and they fail to meet the above asymptotic constraint. From this it follows that whenever we are dealing with a probabilistic regress in which the conditional probabilities are of the usual class, fading foundations will ensue. Indeed, the necessary and sufficient condition for fading foundations is membership of the usual class.

Despite the technicalities we needed to prove it, the result itself is actually very intuitive. If the conditional probabilities in a regress are very close to those corresponding to entailments, then we can only determine the truth value of the target if we know the truth value of the ground. Irrespective of the chain's length, and thus irrespective of whether the ground is very close to the target or is far removed from it, the ground continues to make a contribution, and then the age-old regress problem rears its ugly head. But if the regress contains genuine conditional probabilities, i.e. conditional probabilities that do not resemble implications, then the remainder term goes to zero, and the regress is benign.

Strictly speaking, as we noted in Chapter 3, footnote 29, in the usual class probabilistic support is not needed for convergence. But probabilistic support is important for three reasons. First, we are interested in epistemic justification, and this contains probabilistic support as a necessary element. Whatever it may mean to say that 'A_j justifies A_i', part of its meaning is that $P(A_i|A_j) > P(A_i|\neg A_j)$. Second, we like to see epistemic justification as something that amounts to striking a balance. In justifying our beliefs, we set up a trade-off between the number of reasons that we can handle with our finite minds and the level of accuracy that we want to reach. As we will explain in the next chapter, probabilistic support is needed for such a view of justification as a trade-off. Third and finally, the condition of probabilistic support is needed for the convergence of the networks that we discuss in Chapter 8.

Open Access This chapter is licensed under the terms of the Creative Commons Attribution 4.0 International License (http://creativecommons.org/licenses/by/4.0/), which permits use, sharing, adaptation, distribution and reproduction in any medium or format, as long as you give appropriate credit to the original author(s) and the source, provide a link to the Creative Commons license and indicate if changes were made.

The images or other third party material in this chapter are included in the chapter's Creative Commons license, unless indicated otherwise in a credit line to the material. If material is not included in the chapter's Creative Commons license and your intended use is not permitted by statutory regulation or exceeds the permitted use, you will need to obtain permission directly from the copyright holder.

Chapter 5
Finite Minds

Abstract
Can finite minds encompass an infinite number of beliefs? There is a difference between being able to complete an infinite series and being able to compute its outcome; and justification is more than mere calculation. Yet the number of propositions or beliefs that are needed in order to reach a desired level of justification for the target can be determined without computing an infinite number of terms: only a finite number of reasons are required for any desired level of accuracy. This suggests a view of epistemic justification as a trade-off between the accuracy of the target and the number of reasons taken into consideration.

5.1 Ought-Implies-Can

As in the past, the idea of infinite epistemic chains is still generally regarded as being nonsensical, and often for the same reasons. Scott Aikin has divided the various objections to infinite chains into two main categories: the *ought-implies-can* arguments, which are basically pragmatic in character, and the *conceptual* arguments.[1] In this chapter we deal with the first category; the conceptual arguments we will discuss in the next chapter.

Ought-implies-can arguments in effect contain all the different versions of the notorious finite mind objection, which was already raised by Aristotle. They imply that justifying our beliefs only counts as an obligation in so far as we are capable of doing so. Given our human finitude we cannot complete an

---
[1] Aikin 2011, Chapter 2.

infinite series of inferential justification, hence we are not obliged to perform this task. Aikin distinguishes two kinds of ought-implies-can arguments:

> On the one hand, there are arguments that the *quantity* of beliefs (and inferences) necessary is beyond us (for various reasons). This is the argument from quantitative incapacity. On the other hand, there are arguments that the quality (or kind) of belief necessary to complete the regress appropriately is one we simply cannot have. That is, because some belief in or about the series (and necessary for the series to provide epistemic justification) will be so complex, we cannot have it. And thereby, we cannot maintain the series in a way capable of amounting to epistemic justification. This is the argument from *qualitative* incapacity.[2]

The idea is straightforward enough: because we are mortal and of restricted capacity, we are unable to handle epistemic chains that either contain an infinite number of beliefs or contain some beliefs that are too complicated for us to handle.

But straightforward as it may seem at first sight, the idea is not always clear, and it has not always been expressed in the same way. Even among the philosophers who are most pertinacious in their disapproval of infinite epistemic chains, there is no agreement on this matter. For example, Michael Bergmann, as we have seen, deems it obvious that we cannot have an infinite number of beliefs:

> ...it seems completely clear that none of us has an infinite number of actual beliefs, each of which is based on another ...[3]

Noah Lemos agrees:

> One difficulty with [the option of an infinite chain] is that it seems psychologically impossible for us to have an infinite number of beliefs. If it is psychologically impossible for us to have an infinite number of beliefs, then none of our beliefs can be supported by an infinite evidential chain.[4]

But Richard Fumerton has a different opinion:

> There is nothing absurd in the supposition that people have an infinite number of justified beliefs.[5]

---

[2] Ibid., 52. The same distinction was made by John Williams when he discriminated between an infinite number of beliefs and an infinitely complex belief (Williams 1981).
[3] Bergmann 2007, 23.
[4] Lemos 2007, 48.
[5] Fumerton 2006, 49.

5.1 Ought-Implies-Can

Klein is right that we do have an infinite number of beliefs.[6]

> ...there probably is no difficulty in supposing that people can have an infinite number of beliefs.[7]

This difference of opinion should perhaps not surprise us. After all, as noted earlier, it is entirely unclear how we should count our beliefs. This observation already intimates that knock-down arguments whether we *can* or *cannot* have an infinite number of beliefs are not to be expected.

Peter Klein has defended his infinitism against the finite minds objection by arguing that the objection is based on what he calls the 'Completion Requirement'. According to this requirement, a belief can be justified for a person only if that person has actually completed the process of reasoning to the belief. Such a requirement, says Klein, is against the spirit of infinitism indeed, but it is also unrealistic in that it is too demanding:

> Of course, the infinitist cannot agree to [the Completion Requirement] because to do so would be tantamount to rejecting infinitism. More importantly, the infinitist should not agree because the Completion Argument demands more than what is required to have a justified belief.[8]

Klein regards epistemic justification as being incomplete at heart: it is essentially provisional and can always be further improved. He fleshes out this view by means of two distinctions: the distinction between propositional and doxastic justification, and that between objective and subjective availability. Propositional justification, according to Klein, depends on the objective availability of reasons in an endless chain, where objective availability means that one proposition *is* a reason for another, so that it can be said to justify even if we are not aware of it. Doxastic justification, on the other hand, is parasitic on propositional justification and hinges on an availability that is subjective: a belief q is doxastically justified for an epistemic agent S if there is, in the endless chain of reasons, a reason for q that S can "call on". Although in its entirety the chain can never be subjectively available to S's finite mind, S can take a few steps on the endless path. How many steps S can take, or needs to take in order to reach doxastic justification, all depends on contextual factors:

> Infinitism is committed to an account of *propositional justification* such that a proposition, q, is justified for S *iff* there is an endless series of non-repeating

---

[6] Fumerton 2001, 7.
[7] Fumerton 1995, 140.
[8] Klein 1998, 920.

propositions available to S such that beginning with q, each succeeding member is a reason for the immediately preceding one. It is committed to an account of *doxastic justification* such that a belief is doxastically justified for S *iff* S has engaged in tracing the reasons in virtue of which the proposition q is justified far forward enough to satisfy the contextually determined requirements.[9]

We sympathize with Klein's view, but the previous chapters have made it clear that our position differs in two ways. On the one hand it is *weaker*: where Klein holds that justification requires the objective availability of an infinite chain, we allow that there can be justification even if the chain terminates. In those cases the foundation still exerts some justificatory influence of the target; and just how much justificatory influence it exerts depends on other characteristics of the chain, such as its length and the speed with which the series of conditional probabilities converges. On the other hand, our position is *stronger* than that of Klein: where he denies that infinite chains can be completed, we assert that they can. We only need to construe justification probabilistically and make sure that we are in what we have called 'the usual class', i.e. the domain where the probabilistic support is not too close to entailment.[10]

---

[9] Klein 2007a, 11. We have substituted q for p. *Cf.* Section 1.2.

[10] While some have taken the view that Klein's infinitism can account for propositional but not for doxastic justification, Jonathan Kvanvig has argued it fails on both counts. His argument why it fails for propositional justification goes as follows. In Klein's view, propositional justification either is relative to the total evidence available or is not so relative (where 'available' is interpreted liberally: a reason need not be present in order to be available, but may be only ready to hand). If propositional justification is *not* relative to the total evidence available, then my justification for q might depend on which book I happen to have taken from my shelves: "one source can be the start of an infinite chain of reasons for thinking [q], and the other source the start of an infinite chain for [$\neg q$]" (Kvanvig 2014, 140). If, on the other hand, propositional justification *is* relative to the total evidence available, then scepticism looms. Suppose that evidence E_1 confirms q, that E_2 confirms $\neg q$, and that $E_1 \wedge E_2$ does not confirm q. Let person S_1 have E_1 as evidence, S_2 have E_2, and the infinitist have $E_1 \wedge E_2$. Then, Kvanvig argues, "if propositional justification is relative to total information", none of these three have justification for q or for $\neg q$. Kvanvig's argument rightly points to vagueness in the term 'availability', whether interpreted liberally or strictly. However, his argument seems to presuppose an 'absolute' concept of justification and moreover to equate justification with confirmation. With the relational concept of justification that we proposed in Chapter 2, and with the assumption that confirmation is necessary but not sufficient for justification, there does not seem to be a problem.

5.2 Completion and Computation

On the basis of the previous chapters our answer to the finite mind objection will not come as a surprise. If justification is probabilistically construed, then even the 'Completion Requirement' that Klein rebuts can be met.[11] For then infinite justificatory chains *can* indeed be completed in the sense that they yield a unique and well-defined probability value for the target proposition. And if it is possible to complete infinite chains, the finite mind objection does not arise. Although this answer to the finite mind objection differs from that of Klein, who after all asserts that completion and infinitism are irreconcilable, it does enable us to account for at least two of Klein's intuitions, namely that epistemic justification gradually emerges along the chain and that contextual factors decide at which level of emergence we will decide that 'enough is enough'.[12]

However, Jeremy Gwiazda has argued that this reply to the finite mind objection does not work. As he sees it, we have not *completed* a probabilistic regress, but we have only *computed* its limit.[13] There is a great difference, according to Gwiazda, between calculating the probability value of a target proposition on the one hand and actually giving reasons for that proposition on the other. Gwiazda does not discuss in detail what the differences are, but he might be thinking of a difference in time: while we can calculate the limit of an infinite series in a finite time, we are unable to come up, in a finite time, with an infinite number of reasons. As such, the difference resembles an important distinction that Nicholas Rescher has emphasized, namely between regresses which are time-compressible and those which are not. An example of the former is generated by the Zeno-like thesis 'To *reach* a destination, you must first *reach* the halfway point to it'; an example of the latter is produced by 'To *make a journey* to a destination, you must first *make a journey* to the halfway point to it':

> The first thesis is true — and harmless: that is just how transit from point *A* to point *B* works. But the second is false and, moreover, vicious in rendering any sort of journey impossible. Zeno of Elea notwithstanding, a motion to reach or to cross endlessly many points is perfectly possible. But infinite *journeying*,

---

[11] This point appears to have been missed in Wright 2013.

[12] This is basically the way in which Frederik Herzberg, referring to insights about probabilistic regresses, has replied to the 'new finite mind objection' that was raised by Adam Podlaskowski and Joshua Smith. See Herzberg 2013, 373-374, and Podlaskowski and Smith 2011.

[13] Gwiazda 2010. The same point was made by Matthias Steup (Steup 1989).

with its inherent requirement for explicitly planned and acknowledged transits, is an impossibility. And the reason for this lies not in the impossibility of motion, but in the fact that making a journey to somewhere (as *distinct* from reaching or arriving there) involves deliberation and intentional goal-setting. And since man is a finite being, an infinitude of conscious mental acts is impossible for us. So while that first *structural* regress is harmless, the second regression of infinitely many consciously performed acts is an impossibility.[14]

In the same vein, it could be admitted that we, with our finite minds, are capable of calculating the probability of a target proposition (in the previous chapters we have after all done so), but are incapable of giving an infinite number of reasons for this proposition, since the latter would require an infinity of consciously performed acts. Because epistemological justification is about giving reasons, and not about making calculations, the finite mind objection applies in full force.

A similar reaction to our views has been voiced by Adam Poslaskowski and Joshua Smith.[15] They argue that, although "valuable lessons" can be drawn from our formal results, it is "entirely unclear" that these results meet a basic requirement, namely "providing an account of infinite chains of propositions *qua* reasons made available to agents".[16] Podlaskowski and Smith call this 'the availability problem':

> Given the distinctive emphasis that Peijnenburg, Atkinson, and Herzberg place on calculability, we have doubts about the extent to which (on their account) an infinite chain of propositions can serve as *reasons* that are *available* to an agent. (This is what shall be called the *availability problem* facing the distinctive brand of infinitism under consideration).[17]

> ... it is hard to see, more generally, how the emphasis on calculability yields a notion of *available reason* (or *availability*) that can serve the infinitist's purposes.[18]

Podlaskowski and Smith maintain that our analysis confuses two completely different things, namely being able to compute the probability of a target

---

[14] Rescher 2010, 25. Rescher uses several ways to express the distinction between time-compressible and non-time-compressible regresses; one of them is by saying that the latter need pre-conditions whereas the former only has co-conditions (ibid., 55).
[15] Podlaskowski and Smith 2014.
[16] Ibid., 212.
[17] Ibid., 214.
[18] Ibid., 215.

proposition on the one hand and having available reasons for this proposition on the other. They blame us for assuming that, "since mathematical means exist with which an agent can decide the probability of any proposition being true (even if it belongs to an infinite series), all the members of an infinite chain of reasons must thereby be *available* (as reasons) to an epistemic agent".[19] Like Gwiazda, they stress the difference between determining the probability of a target q and showing that something is a reason for q:

> deciding the probability of any given proposition ... even if there are infinite chains of propositions ... is still a far cry from showing that, as a matter of principle, each proposition in a chain of of propositions is one that can serve as a *reason* for another proposition in that chain, and do so in the right order. It appears that two dispositions have been conflated: those to make a certain sort of calculation, and those to accept any given proposition as reason for another proposition. ... [A] demonstration that finite agents can actually calculate the probability of a proposition's truth — even if it belongs to an infinite chain of reasons — does not thereby show that each reason is equally *available* to a finite agent.[20]

The observation of Gwiazda and Podlaskowski and Smith that computing and completing reflect two different dispositions is fair enough. However, as we will explain in the next section, in epistemic justification we draw on both. In this sense, justification resembles logic: there, too, we draw on an abstract, normative dimension concerning how one *ought* to reason, and a concrete, descriptive dimension concerning how one reasons *in fact*.[21] Together the two dimensions suggest a view of justification as a trade-off between the accuracy of the target proposition and the capacity of our mental housekeeping.

5.3 Probabilistic Justification as a Trade-Off

Rescher is of course right that a time-compressible regress is different from a regress that is not time-compressible. And Gwiazda and Podlaskowski and Smith are right that making a calculation is not the same as giving a proposition as reason for another proposition. The skill to compute the value of the

---

[19] Ibid., 215.

[20] Ibid., 216. Michael Rescorla's complaint that our approach falls prey to 'hyperintellectualism' expresses a similar sentiment (Rescorla 2014).

[21] Van Benthem 2014, 2015.

target on the basis of a probabilistic epistemic chain indeed differs from the capacity to have the propositions in the chain available as reasons.

However, these two faculties are not disjunct, as the above authors seem to think.[22] Especially when it comes to epistemic justification, of which probabilistic support is an essential part, these faculties are closely and essentially connected. A justificatory regress is not just any old regress; it is a regress about *reasoning*, in this case reasoning that involves how a proposition or belief is probabilistically justified by another. This means that the actual process of 'giving probabilistic reasons' is to a certain extent subjected to the rules of the probability calculus, just as the actual process of 'giving deductive reasons' is to a certain extent subject to the rules of deductive logic. The aversion of Gwiazda and others to using calculations in the context of giving reasons might be exacerbated by the idea that this necessarily involves processing an infinite number of terms. That idea, although understandable, is however mistaken, and betrays a misconstrual of our view.

We have argued that, whenever we give a reason, A_i, for a target q, the significance of A_i as a reason depends on how much probabilistic support it gives to q. The latter in turn depends on how much A_i's support for q deviates from the 'final' support, i.e. the support that q would receive from the entire justificatory chain of which A_i is a member. And how much support q receives from the entire infinite chain depends on the chain's character, i.e. on the values of its conditional probabilities together with the value of the unconditional probability of the ground, p. While the conditional probabilities come from experiments, the unconditional probability of the ground is unknown.[23] The longer the chain, the smaller the contribution from the ground, and when the chain is infinitely long, the contribution from the ground to the target vanishes completely, leaving all the justificatory support to come from the combined conditional probabilities.

The view could be easily misunderstood. It does *not* imply that 'giving reasons' depends on 'making calculations' in the sense that we first have to calculate the limit of a probabilistic regress before we can know what our reason is worth; computing the limit is *not* necessary for weighing the quality of our actual reasons. Rather, the structure of the probabilistic justificatory

---

[22] Recall the claim of Podlaskowski and Smith that it is "entirely unclear" what formal calculation means for "propositions *qua* reasons made available to agents" (Podlaskowski and Smith 2014, 212).

[23] In Chapter 8, Section 8.5, we will come back to the status of the conditional probabilities. In particular, we consider the situation in which they are not given, but are themselves in need of justification. As we will explain, a network is then created with a remarkable structure that resembles a Mandelbrot fractal.

5.3 Probabilistic Justification as a Trade-Off

chain is such that it enables us to say *how many reasons* we need to call on in order to approach the probability of the target to a satisfactory level. To do that, we do not need to know the length of the chain; we need not even know whether it is finite or infinite. Nor do we have to know the probability of the ground. The only thing we need are the values of a certain number of conditional probabilities (sometimes more, sometimes less, depending on the speed of the convergence) that suffice to take us to within a desired level of accuracy with respect to the true, but unknown probability of the target. Once we are there, we can safely ignore the rest of the chain — such is the lesson of fading foundations.

An example might help to understand the point. Imagine I have a reason A_1 for my belief q, and know the two relevant conditional probabilities, $P(q|A_1)$ and $P(q|\neg A_1)$. Suppose I am unable or unwilling to back up A_1 by a further reason, and therefore want to cut off the chain here. We have seen that knowing the conditional probabilities is in general not enough to know the value of $P(q)$; especially with short chains like the one at hand it is indispensable that we also know the unconditional probability $P(p)$. Even if I have no clue what the value of the latter is, I do know that it cannot be greater than one and cannot be smaller than zero. I now consider these two extremal cases, i.e. where $P(p) = 1$ and where $P(p) = 0$, and I find that in the first case $P(q) = x$ and in the latter case $P(q) = y$. The condition of probabilistic support now guarantees that the real value of $P(q)$ lies in the interval between x and y, *no matter how many further* A_n we take into consideration. What is more, the condition ensures that with every reason we add, the interval will *become smaller*, making the value of $P(q)$ more precise with each step. This applies both in the uniform situation, where the conditional probabilities are all the same, and in the nonuniform case, where they are different.

As a result, I can determine how many reasons I need to have in order to approach the true probability of the target q within an error margin of, for example, 1%. If this number of reasons happens to be too large to fit into my finite mind, then I will have to relax the level, and be content with a degree of justification that is further away from the true probability of the target. But if the number of reasons is rather small, so that they all fit in my finite mind (although perhaps not in that of my four-year-old daughter), then I can always tighten up the satisfaction level, and come closer to the target's true probability. Epistemic justification thus boils down to striking a balance. In acting as responsible epistemic agents, we are instigating a trade-off between the number of reasons that we can handle and the level of accuracy that we want to reach. If we are unable or unwilling to manage a large number of reasons, we have to pay in terms of a lack of precision and hence of trustworthiness. Taking the short route thus comes at a price, but in situations where precision is not important, we can take it easy and should do so on pain of exerting ourselves unnecessarily.

Let us spell out this idea of a trade-off more fully and more formally. Assume a finite chain to consist of five propositions, the target proposition q, the intermediate propositions A_1 to A_3, and the ground A_4:

$$P(q) = \beta_0 + \gamma_0\beta_1 + \gamma_0\gamma_1\beta_2 + \gamma_0\gamma_1\gamma_2\beta_3 + \gamma_0\gamma_1\gamma_2\gamma_3 P(A_4). \quad (5.1)$$

As we explained in Sections 3.5 and 3.6, the right-hand side consists of two terms. The first term is the sum of the conditional probabilities,

$$\beta_0 + \gamma_0\beta_1 + \gamma_0\gamma_1\beta_2 + \gamma_0\gamma_1\gamma_2\beta_3,$$

and the second is the remainder term,

$$\gamma_0\gamma_1\gamma_2\gamma_3 P(A_4).$$

This remainder term is a product of two factors, $\gamma_0\gamma_1\gamma_2\gamma_3$ and $P(A_4)$. Since we suppose the conditional probabilities to be known, there is only one probability that we need to know in order to compute $P(q)$. This is $P(A_4)$, i.e. the unconditional probability of the ground. If we did know $P(A_4)$, then we would know $P(q)$.

However, suppose we have no clue as to the value of $P(A_4)$. What to do? Because of the condition of probabilistic support, (2.1), all the γ_n are positive, which means that every term in (5.1) is positive too. Therefore the smallest value that $P(q)$ could have, given the conditional probabilities, is obtained by giving $P(A_4)$ the minimum value that it could have, which is zero, leaving only

$$\beta_0 + \gamma_0\beta_1 + \gamma_0\gamma_1\beta_2 + \gamma_0\gamma_1\gamma_2\beta_3. \quad (5.2)$$

On the other hand, the largest value that $P(q)$ could have is obtained by giving $P(A_4)$ the maximum value that it could have, which is one, yielding

$$\beta_0 + \gamma_0\beta_1 + \gamma_0\gamma_1\beta_2 + \gamma_0\gamma_1\gamma_2\beta_3 + \gamma_0\gamma_1\gamma_2\gamma_3. \quad (5.3)$$

We know that the value of $P(q)$ must lie somewhere between the two extremes (5.2) and (5.3). If we were to assume the value of $P(q)$ to be one extreme, for example (5.2), then we would be sure that our error could not be larger than the difference between the maximum, (5.3), and the minimum, (5.2), namely $\gamma_0\gamma_1\gamma_2\gamma_3$.

Now imagine that the error term $\gamma_0\gamma_1\gamma_2\gamma_3$ turns out to be, for example, only 1% of the minimum value (5.2). And suppose further that we proclaim ourselves satisfied with a value that deviates by no more than 1% from the true

5.3 Probabilistic Justification as a Trade-Off 111

value of $P(q)$. Then we need go no further in inquiring as to any support that the ground, A_4, might have from some other proposition. This is because any extension of the chain, obtained by adding a proposition, A_5, that supports the erstwhile ground A_4 would only *increase* the minimum (5.2) to

$$\beta_0 + \gamma_0\beta_1 + \gamma_0\gamma_1\beta_2 + \gamma_0\gamma_1\gamma_2\beta_3 + \gamma_0\gamma_1\gamma_2\gamma_3\beta_4,$$

and *decrease* the error to $\gamma_0\gamma_1\gamma_2\gamma_3\gamma_4$ (this *is* smaller, because the extra factor, γ_4, is less than one). This is precisely what fading foundations imply. So in this case we know exactly how many reasons we need in order to approach the true value of the target to a level that satisfies us. If we are content with a value that deviates no more than 1% from the true value of $P(q)$, then we require no more than four reasons for q, namely A_1 to A_4. And if our mind is big enough to store these four reasons, then we have accomplished our task: we have justified q to a satisfactory level, staying neatly within the limitations of our finite mind. Note that we have performed our task without knowing the true value of $P(q)$ or that of $P(A_4)$.

What to do when the error term $\gamma_0\gamma_1\gamma_2\gamma_3$ turns out to be very big, for example 90% of the minimum value (5.2)? How should we proceed now? If our level of required accuracy is still 1%, then there is not much that we can do in this case. We might sadly conclude is that there is much uncertainty, due to the fact that the justificatory influence of the unknown $P(A_4)$ on $P(q)$ is very great, but that is as far as we can get. For the four reasons that we can avail ourselves of, A_1 to A_4, are of little help: jointly they bring us to a point where the deviation from the true value of $P(q)$ may be as great as 90%.

However, let us now make the finite chain considerably longer. Rather than assuming that there are four reasons for q, let us suppose that there are one hundred:

$$P(q) = \beta_0 + \gamma_0\beta_1 + \gamma_0\gamma_1\beta_2 + \ldots + \gamma_0\gamma_1\ldots\gamma_{m-1}\beta_m + \gamma_0\gamma_1\ldots\gamma_m P(A_{m+1}), \quad (5.4)$$

where $m = 99$. It is unlikely that I can store all these reasons in my finite mind, so I decide to cut off chain (5.4) at number seven, making a provisional stop at proposition A_6. So I get:

$$P(q) = \beta_0 + \gamma_0\beta_1 + \gamma_0\gamma_1\beta_2 + \ldots + \gamma_0\gamma_1\gamma_2\gamma_3\gamma_4\beta_5 + \gamma_0\gamma_1\gamma_2\gamma_3\gamma_4\gamma_5 P(A_6). \quad (5.5)$$

In formula (5.5) I can only compute $P(q)$ if I know $P(A_6)$. Since I have no idea as to the value of the latter, I apply the same reasoning as above. That is, I first recall that the value of $P(q)$ must lie between two extremes. The one extreme is obtained by putting the unknown $P(A_6)$ equal to zero. The other extreme is obtained by putting it equal to one. Suppose that I adopt the first extreme, $P(A_6) = 0$, so my estimation of $P(q)$ is that it has its minimum value. I know that my error in making this estimation cannot be larger than the difference between this minimum value and the maximum value of $P(q)$, obtained by putting $P(A_6) = 1$. The difference itself is given by our error term, which in this case is $\gamma_0 \gamma_1 \gamma_2 \gamma_3 \gamma_4 \gamma_5$.

Now suppose that the error term $\gamma_0 \gamma_1 \gamma_2 \gamma_3 \gamma_4 \gamma_5$ is only 1% of the minimum value of $P(q)$. And suppose again that I am satisfied with an accuracy that deviates no more than 1% from the true value of $P(q)$. If I am capable of storing six reasons in my head, then I am done. In particular, I do not have to go on and find a justification for A_6. A we have seen, the reason for this lies in the fact that, as the chain lengthens, the minimum value of $P(q)$ increases and the maximum value decreases — which is a direct consequence of the condition of probabilistic support. This condition implies that any extension of the chain would only make the minimum value of $P(q)$ greater, and thus would make the error term itself smaller.[24] Consequently, adding a proposition to chain (5.5), for example proposition A_7, would bring us closer to the true value of $P(q)$; and since we are already satisfied with our level of approximation, there is no need to engage in this project. We have in fact reached the point where 'enough is enough', and this expression now has a very precise meaning. For any justificatory chain, I can first define the level of accuracy within which I want to approach the true value of $P(q)$, and I can then determine how many reasons I need to reach this level. In order to perform these tasks, I need not know the value of $P(q)$, nor that of $P(A_6)$, nor that of any other ground. More importantly, I can blissfully neglect the rest of the chain. For not only is it so that I am within the desired 1% of the true probability value, it is also the case that calling on any further reason will only bring me closer to that true value. As the chain gets longer, the remainder term gets smaller (in accordance with fading foundations) and the sum of the conditional probabilities gets larger (in accordance with the condition of probabilistic support). So as m gets bigger, the value of the sum of the conditional probabilities increases monotonically, whereas the remainder term decreases monotonically. Therefore, if we are already satisfied with 1%, any extension of the chain will bring us still closer to the true value of $P(q)$; there is thus no need to call on more reasons than the six reasons that we have (subjectively) available.

But now suppose that the error term $\gamma_0 \gamma_1 \gamma_2 \gamma_3 \gamma_4 \gamma_5$ in formula (5.5) still greatly differs from the minimum value of $P(q)$; let us now say by 80%. The

---

[24] See Appendix A.2 for the proof.

5.3 Probabilistic Justification as a Trade-Off

situation is not the same as it was in the case (5.1). Since Eq.(5.5) is part of a larger chain we have the option to go on, and to look for the justification of A_6 in terms of A_7; after that, we can go further and justify A_7 in terms of A_8, and so on. The more propositions we add, the more we lengthen our chain, and the smaller will be the difference between (5.4) and the minimum value of $P(q)$. We are now able to reduce the error to less than 80% of the true value of $P(q)$. However, there is a price to pay. In getting closer and closer to the real value of $P(q)$, we are calling on more and more reasons, and our finite minds have to accommodate each and every extra reason that we call on. It could happen that our minds lack the capacity to take in all the reasons that our level of accuracy requires. In that case the only option left open for us is to relax the accuracy level to a degree where it corresponds to a number of reasons that *can* be housed in our heads. We are committed to a trade-off: we simply cannot have our cake and eat it too.[25]

What we have said above is of course not restricted to finite chains such as (5.1), (5.4), and (5.5). The reasoning about error terms works just as well with an infinite chain as with a finite chain. In both cases we can work out, in a finite number of steps, how many terms we need to reach a particular, pragmatically determined level of accuracy. If it turns out that our level of accuracy requires more reasons than we can accommodate, then we are living

---

[25] Whether a particular number of reasons can or cannot be housed in our heads might depend not just on size or on capacities, but also on other factors. Linda Zagzebski has distinguished between two kinds of epistemic reasons for believing a proposition q: theoretical reasons, which are third personal and "connect facts about the world with the truth of $[q]$", and deliberative reasons, which are first personal and "connect me to getting the truth of $[q]$" (Zagzebski 2014, 244). Even if, impossibly, we were able to complete our search for theoretical reasons, that would still leave us with the second problem that what we call 'reasons' may not indicate the truth: "We would still need trust that there is *any* connection between what we think are the theoretical reasons and the truth" (ibid., 250). Zagzebski argues that this second problem can only be solved by calling on a deliberative reason with a special status, viz. epistemic self-trust, which ends our urge to search for further theoretical or deliberative reasons. It is not excluded that Zagzebski's epistemic self-trust might be a factor in the process of trading-off. Other possible factors might perhaps be the localist considerations of Adam Leite, or the "plausibility considerations" that Ted Poston mentions in support of his claim that "there is more to epistemic justification than can be expressed in any reasoning session" (Leite 2005; Poston 2014, 182-183). We expect that our trade-off can also be combined with Andrew Norman's "dialectical equilibrium" (Norman 1997, 487) and Michael Rescorla's "dialectical egalitarianism" (Rescorla 2009), although we are not sure if the authors themselves would agree.

beyond our means. We then should either work harder and try to create more space in our finite minds, or become more modest and lower our desire for accuracy. All this can be done without having to call on, or even to calculate, all the terms in a (finite or an infinite) series.

The two tables below illustrate the idea. In the first, the conditional probabilities, α and β, have the values 0.99 and 0.04 respectively; in the second, they are 0.95 and 0.45. 'Maximum $P(q)$' and 'Minimum $P(q)$' refer to the values that $P(q)$ has when $P(p)$ is one or zero, respectively.

Table 5.1 Extremal values of $P(q)$ when $\alpha = 0.99$ and $\beta = 0.04$.

Number of A_n	1	2	5	10	15	25	50	100	∞
Minimum $P(q)$	.078	.114	.212	.345	.448	.589	.742	.796	.8
Maximum $P(q)$	.981	.971	.947	.914	.888	.853	.815	.801	.8

Table 5.2 Extremal values of $P(q)$ when $\alpha = 0.95$ and $\beta = 0.45$.

Number of A_n	1	2	3	4	5	6	8	10	∞
Minimum $P(q)$	.675	.788	.844	.872	.886	.893	.8982	.8996	.9
Maximum $P(q)$	.925	.913	.906	.904	.902	.901	.9002	.9000	.9

In the first table one needs more than fifty intermediate reasons A_n to ensure that the difference between the maximum and the minimum of $P(q)$ is relatively small, whereas in the second table a similar uncertainty is already reached after a mere three reasons A_n. There the situation is much more amenable. Justification as a form of trade-off sheds light on the difference between propositional and doxastic justification that we discussed in 4.2. Some scholars appear to be of the opinion that propositional and doxastic justification can never be combined, since the former is abstract and infinite, while the latter is concrete and finite by definition.

Others have however argued that doxastic justification is parasitic on propositional justification, and that the context determines when exactly it comes to an end. Our considerations in this section clarify the latter position, and they make clear how this contextualism can be interpreted.

5.4 Carl the Calculator

When commenting on our approach, Podlaskowski and Smith write that "care must be taken when assessing the significance of these formal results".[26] Of course we agree, and it can be added that the same applies to assessing results that are not formal: whenever we informally discuss reasoning, or justification, or probability, we must take care what we say. For example, as we have seen, it is incorrect to say that an infinite probabilistic regress yields zero for the target, or that knowing the value of the target requires knowing the value of a basic belief. Intuitive as these claims might be, they are incorrect as they stand.

The difference of opinion between Podlaskowski and Smith and us, if there is one, concerns the relation between the ability to calculate and the ability to give reasons. As we explained in the previous section, we believe that epistemic justification involves both. Podlaskowski and Smith seem however to interpret us differently, thinking that for us having the mathematical ability to calculate is sufficient for having justification. This is for instance the message from their instructive example about Carl, who is a real pundit when it comes to calculating probabilities, but who cannot understand the meaning of reasons:

> [I]magine Carl, whose impressive talent in calculating conditional probabilities is strangely at odds with his ability to grasp various concepts. Carl has no problem solving all manner of complex equations, including those involving conditional probabilities (such as Peijnenburg, Atkinson, and Herzberg provide). Yet, there are various concepts which he is entirely incapable of grasping, some of which might feature in reasons whose probabilities of being true are conditional on other reasons. Suppose that Carl is given two lists, an infinite list of conditional probability assignments and an infinite list of reasons. Unbeknownst to Carl, the two lists correspond perfectly: the list of probabilities is meant to capture the probability of each reason being true, conditional on its predecessor. Moreover, some of the members of the list of reasons are comprised of those concepts that Carl is incapable of grasping. Even if Carl were capable of working through some infinite list of reasons, at some point on the list at hand, Carl would fail to comprehend the concepts deployed. But he would have *no problem* doing the corresponding calculations. Does merely calculating the probability of the chain make Carl justified in holding any of those beliefs, when Carl is *incapable* of understanding the concepts on which those beliefs depend? Surely not. If an agent *cannot* understand some of the

---
[26] Podlaskowski and Smith 2014, 212.

reasons in the infinite chain, it is difficult to see how those reasons can do any justificatory work for him.[27]

Podlaskowski and Smith suggest that, according to us, Carl has justified his beliefs. This is however not so: for us, as for Podlaskowski and Smith, Carl fails to justify. Our view is not that calculation implies justification, but that justification implies a certain amount of 'calculation'. Of course we realize that people often put forward probabilistic reasons for their beliefs without knowing anything about the probability calculus. As epistemologists who want to take the concept of probabilistic reasoning seriously, however, we believe that a minimum of adjustment to the probability calculus seems to be required, even if it is only in a rational reconstruction.

Podlaskowski and Smith seem to have anticipated this response when they write:

> One might ... suspect that we have crafted the Carl case too narrowly, and that it misses some important aspect of what mathematical analyses of probabilistic regresses are supposed to be doing.[28]

However, they then suggest that our response requires a new notion of 'available reason' which cannot be developed within our approach:

> Perhaps there is a notion of *available reason* that can supplement the project of Peijnenburg et al. that avoids the problems raised by the Carl case. The problem with successfully developing such a response, however, is that it is entirely unclear what sort of notion they could use, given their emphasis on calculability. To see this, consider the spectrum of possible views. On one end, the notion of *availability* drops out. This end of the spectrum has the unfortunate consequence that the view collapses into maintaining that a belief is justified for a person when there merely exists an infinite, non-repeating chain of reasons that makes the belief probable. ... On the other end of the spectrum, one might hold a very strong notion of *availability*, according to which it is required that one *actually believe* a reason for it to be available. But this is far too strong, as it runs face-first into the original finite minds objection to infinitism. ... One lesson to draw from the Carl case is that moving a brand of infinitism beyond Klein's middle ground on the notion of *availability* proves seriously problematic ....[29]

Here Podlaskowski and Smith write as if there are only two possibilities: either we merely calculate, and then no reason qua reason is available, or

---

[27] Ibid., 216.
[28] Ibid., 217.
[29] Ibid.

5.4 Carl the Calculator

we hold on to a strong notion of availablity, but then we run into the finite mind objection. Our remarks in the previous section provide us with a notion of availability that avoids the two extremes that Podlaskowski and Smith present. Often it is enough that only a few reasons are available in order to draw conclusions that go far beyond what is implied by these available reasons themselves. If the reasons in question bring us close enough to the true value of the target, then the phenomenon of fading foundations tells us that we can ignore the rest of the chain. If, on the other hand, the reasons do not bring us within a desired level of accuracy, then we will have to achieve a balance between the number of reasons that we can handle and the degree to which we can approach the final value of the target. Thanks to the condition of probabilistic support we can determine how many reasons we need in order to conclude that the rest of the chain is irrelevant. In this chapter we have explained the idea in quantitative terms, but it can quite easily be grasped in an intuitive and qualitative way.

Open Access This chapter is licensed under the terms of the Creative Commons Attribution 4.0 International License (http://creativecommons.org/licenses/by/4.0/), which permits use, sharing, adaptation, distribution and reproduction in any medium or format, as long as you give appropriate credit to the original author(s) and the source, provide a link to the Creative Commons license and indicate if changes were made.

The images or other third party material in this chapter are included in the chapter's Creative Commons license, unless indicated otherwise in a credit line to the material. If material is not included in the chapter's Creative Commons license and your intended use is not permitted by statutory regulation or exceeds the permitted use, you will need to obtain permission directly from the copyright holder.

Chapter 6
Conceptual Objections

Abstract

There are two conceptual objections to the idea of justification by an infinite regress. First, there is no ground from which the justification can originate. Second, if a regress could justify a proposition, another regress could be found to justify its negation. We show that both objections are pertinent to a regress of entailments, but fail for a probabilistic regress. However, the core notion of such a regress, i.e. probabilistic support, leaves something to be desired: it is not sufficient for justification, so something has to be added. A threshold condition? A closure requirement? Both? Furthermore, the notion is said to have inherent problems, involving symmetry and nontransitivity.

6.1 The No Starting Point Objection

In the previous chapter we discussed the main pragmatic argument against justification by infinite chains, known as the finite mind objection. Perhaps even more serious, however, are the conceptual objections. They aim to show that even creatures with an infinite lifespan or with a mind that can handle infinitely long or complex chains will run into problems, because the very idea of justification is at odds with a chain of infinite length:

> conceptual arguments ... appeal ... to the incompatibility of the concept of epistemic justification and infinite series of support.[1]

Two conceptual objections in particular are often discussed. According to the first, *no* proposition can ever be justified by an infinite regress, since in such

---
[1] Aikin 2011, 51.

a regress justification is for ever put off and never materialized. This is the much raised *no starting point objection*, as Peter Klein has called it, which is based on the fact that an infinite chain is bereft of a source or a foundation from which the justification could spring.[2] The second conceptual objection goes beyond the first one, spelling out what would happen if the no starting point objection did not apply. If, *per impossibile*, a particular proposition q were justified by an infinite chain, then it can be demonstrated that *all* propositions could be justified in that manner, including the negation of q. This objection is known as the *reductio argument*, and it has been raised in different forms, notably by John Pollock, Tim Oakley, James Cornman, Richard Foley, and John Post.[3]

In the present section and in the next one we discuss the no starting point objection. We shall argue that a starting point is not needed if the regress is probabilistic — a conclusion which follows from the preceding chapters. In Sections 6.3 and 6.4 we shall deal with the reductio argument, showing that this objection, too, fails for a probabilistic regress. In the final section, 6.6, we note that the concept which is central to a probabilistic regress, viz. probabilistic support, is itself prone to problems. We elaborate on two properties of probabilistic support that are allegedly problematic for the concept of justification, namely that probabilistic support is symmetric and that it lacks transitivity.

The no starting point objection asserts that justification can never be created by inferences alone. The reason is that an infinite inferential chain blocks *ab initio* the possibility of justification. The only way to generate justification is by having a starting point, i.e. a proposition or a belief that is itself non-inferentially justified. Aikin phrases the objection as follows:

> ...if reasons go on to infinity, then as far as the series goes, there will always be a further belief necessary for all the preceding beliefs to be justified. If there is no end to the chain of beliefs, then there is no justification for that chain to inherit in the first place.[4]

The no starting point objection exploits the fact that in an infinite regress justification seems to be indefinitely postponed and never cashed out. It is as if we are given a cheque with which we go to a bank teller, who gives us a new cheque and directs us to another bank teller, who hands us a third

---

[2] Klein 2000, 204. *Cf.* Laurence Bonjour: "The result ...would be that justification could never get started and hence that no belief would be genuinely justified" (Bonjour 1976, 282).
[3] Pollock 1974, 29; Oakley 1976; Cornman 1977; Foley 1978; Post 1980.
[4] Aikin 2011, 52.

6.1 The No Starting Point Objection

cheque, instructing us to go to yet another bank teller, and so on and so forth. Never do we encounter a bank teller who actually converts our current cheque into bars of gold.

Like the finite mind objection, this objection too has a long history, going back indeed to Aristotle. Aikin recalls some of the latest versions:

> William Alston captures the argument as follows: If there is a branch [of mediately justified beliefs] with no terminus, that means that no matter how far we extend the branch the last element is still a belief that is mediately justified if at all. Thus, as far as this structure goes, whenever we stop adding elements we have still not shown that the relevant necessary condition for mediate justification of the original belief is satisfied. Thus the structure does not exhibit the original belief as mediately justified [Alston 1986, 82].
>
> Henry Johnstone captures the thought: 'X infinitely postponed is not an X' since the series of postponements shortly becomes 'inane stammering' [Johnstone 1996, 96].
>
> Romane Clark notes that such a series will produce only 'conditional justification' [Clarke 1988, 373], and Timo Kajamies calls such support 'incurably conditional' [Kajamies 2009, 532].
>
> The same kind of thought can be captured with an analogy. Take the one R.J. Hankinson uses in his commentary on Sextus: 'Consider a train of infinite length, in which each carriage moves because the one in front of it moves. Even supposing that fact is an adequate explanation for *the movement of each carriage*, one is tempted to say, in the absence of a locomotive, that one still has no explanation for *the motion as a whole. And that metaphor might aptly be transferred to the case of justification in general*' [Hankinson 1995, 189].[5]

In the same vein, Carl Ginet writes:

> Inference cannot *originate* justification, it can only *transfer* it from premises to conclusion. And so it cannot be that, if there actually occurs justification, it is all inferential ... [T]here can be no justification to be transferred unless ultimately something else, something other than the inferential relation, does create justification.[6]

Ginet cites Jonathan Dancy, who phrases the no starting point objection as follows:

> Justification by inference is conditional justification only; [when we justify A by inferring it from B and C] A's justification is conditional upon the justification of B and C. But if all justification is conditional in this sense, then nothing can be shown to be actually, non-conditionally justified.[7]

---

[5] Ibid., 53 – misspellings corrected.
[6] Ginet 2005, 148.
[7] Dancy 1985, 55.

The no starting point objection is also at the heart of Richard Fumerton's "conceptual regress argument" against justificatory chains. On several occasions Fumerton has distinguished between two "regress arguments" in support of foundationalism: the epistemic and the conceptual regress argument.[8] The first boils down to the finite mind objection against infinite chains. It states that "having a justified belief would entail having an infinite number of different justified beliefs" while in fact "finite minds cannot complete an infinite chain of reasoning".[9] In the previous chapter we have explained why we think that this objection does not succeed. The conceptual regress argument, on the other hand, appears to be a rewording of the no starting point objection. Fumerton calls it "quite different" from the epistemic regress argument, and "more fundamental".[10] It states that an infinite justificatory chain is vicious because we can only understand the concept of inferential justification if we accept that of *non*inferential justification:

> [I]f we are building the principle of inferential justification into an analysis of the very concept of justification, we have a more fundamental vicious *conceptual* regress to end. We need the concept of a noninferentially justified belief not only to end the epistemic regress but to provide a conceptual building block upon which we can understand all other sorts of justification. I would argue that the concept of noninferential justification is needed ... in order to *understand* other sorts of justification ....[11]

In other words, the very idea of inferential justification does not make sense without assuming justification that is noninferential, or, as Fumerton formulates it later, the concept of inferential justification is "parasitic" on that of noninferential justification:

> To *complete* our analysis of justification we will need a base clause — we will need a condition sufficient for at least one sort of justification the understanding of which does not already presuppose our understanding the concept of justification. But that sort of justification is just what is meant by noninferential justification (justification that is *not* inferential). Our concept of inferential justification is parasitic upon our concept of noninferential justification. It doesn't follow, of course, that anything falls under the concept. But if nothing does, then there is no inferential justification either ...[12]

---

[8] Fumerton 1995, Chapter 3; Fumerton 2004; Fumerton and Hasan 2010; Fumerton 2014.
[9] Fumerton 1995, 89; 2004, 150; 2006, 40; 2014, 76.
[10] Fumerton 1995, 89; 2014, 76.
[11] Fumerton 1995, 89.
[12] Fumerton 2014, 76.

6.1 The No Starting Point Objection

Not surprisingly, Fumerton's response to his conceptual regress argument echos the standard reply to the no starting point objection: the only way to inject justification into an inferential chain is to assume a source from which the justification springs. Without such a source, the very concept of inferential justification becomes unintelligible or even absurd, 'inane stammering' as Henry Johnstone would have it.

A particularly interesting and generalized version of the no starting objection has been put forward by Carl Gillett.[13] The problem with an infinite chain of reasons, Gillett says, does not lie in its epistemological character as such, but is more general: it has to do with its general metaphysical structure, which it shares with many vicious regresses outside epistemology. This structure is such that the relevant dependent property (which in the epistemological case is 'being justified') cannot be produced, because there is a relation of dependence, what Gillett calls the 'in virtue of' relation. If a proposition q is justified in virtue of A_1 being justified, which in turn is justified in virtue of A_2 being justified, then it is notoriously unclear how any of the propositions could be justified. Making the chain longer is of course no solution, for irrespective of the number of propositions we add, each proposition will only be justified because of another proposition. Thus, Gillett concludes, there is no number of propositions that can be added "that will suffice for any of its dependent properties to feed back to any members of the chain".[14] According to the 'Structural Objection', as Gillett has dubbed his particular version of the argument, the very structure of the epistemic regress prevents justification from arising.

In none of these different formulations of the no starting point objection is it made clear what exactly is meant by epistemic justification. When for example Dancy complains that, "if all justification is conditional ... then nothing can be shown to be actually, non-conditionally justified", it is not clear what he means by 'conditional' and 'non-conditional', since it remains open whether he sees justification as for example entailment or as involving probabilistic support (see Chapter 2). In the first case, his talk about conditional and non-conditional justification would refer to the difference between if-then statements and categorical statements; in the second case, it pertains to the difference between conditional and unconditional probability statements. The distinction is however vital in a discussion of the no starting point objection. For while the objection applies to justification as entailment, as applied to justification as probabilistic support it backfires completely. This result

---

[13] Gillett 2003.
[14] Ibid., 713.

was already intimated in the previous chapter, but we will explain it further in the next section.

6.2 A Probabilistic Regress Needs No Starting Point

It is not difficult to see why the no starting point objection applies if justification is interpreted as a kind of entailment. Consider the finite chain

$$A_0 \longleftarrow A_1 \longleftarrow A_2 \longleftarrow A_3 \longleftarrow \ldots \longleftarrow A_m \longleftarrow A_{m+1} \qquad (6.1)$$

where the arrow represents entailment, where A_0 does duty for the target, q, and where A_{m+1} stands for the foundation or ground. Then of course the only way to know for sure if A_0 is true is by knowing that A_{m+1} is true. In the words of Aikin: "Conceptual arguments start from the deep, and I think right, intuition that epistemic justification should be pursuant of the truth".[15] But if we are ignorant of the truth or falsity of the ground, A_{m+1}, we are groping in the dark about the truth value of A_0. When we make chain (6.1) infinite, so that it looks like:

$$A_0 \longleftarrow A_1 \longleftarrow A_2 \longleftarrow A_3 \longleftarrow A_4 \longleftarrow \ldots \qquad (6.2)$$

then the matter is worse: since there is no initiating A_{m+1}, there is no truth value that is preserved in the first place. For the only way in which the target can be justified is by receiving the property from its neighbour, which received it from its neighbour, and so on. If there is no origin from which the property is handed down, there is nothing to receive, so the no starting point objection applies in full force.

Things are very different when justification is interpreted probabilistically. Applied to a probabilistic chain, the no starting point objection means that the target can only be justified by a chain of conditional probabilities if we know the unconditional probability of the ground. That is, in order to know $P(A_0)$, we need to know not only all the $P(A_j|A_{j+1})$ and $P(A_j|\neg A_{j+1})$, but also the unconditional probability $P(A_{m+1})$. But if the chain is infinitely long, there is no A_{m+1}, and thus there is no probability of A_{m+1} that can be known in the first place. As a result, the no starting objection concludes, there is no way to know the value of $P(A_0)$.

In the previous chapters we have seen why this conclusion does not follow. In all but the exceptional cases, the value of $P(A_0)$ can be determined

---

[15] Aikin 2011, 51.

6.2 A Probabilistic Regress Needs No Starting Point

without having to know the value of some $P(A_{m+1})$. In fact, as we saw in Chapter 5, in many cases we do not even need to know the values of all the conditional probabilities; once we have fixed a particular level of accuracy with which we are satisfied, we can decide how many conditional probabilities we need to know in order to attain that accuracy. If the number of conditional probabilities turns out to be too big to handle, then we must adjust the accuracy level and make do with an approximation of the target's true value with an error margin that is somewhat bigger than we had initially envisaged. So while the no starting point objection implies that in an infinite regress the value of $P(A_0)$ either goes to zero or remains unknown, neither of these two options actually obtains when the probabilistic regress is in the usual class.

John Pollock has trenchantly criticized what he calls "the nebula theory" of justification: never can an infinite chain justify a target, since the chain's ground is for all future time hidden in "a nebula".[16] Pollock would be right that this is an insuperable problem so long as we are speaking about a regress of entailments; but in a probabilistic regress the difficulty does not arise at all. For all we care A_∞ may forever lie hidden in nebulae, in a probabilistic regress that does not matter since A_∞ is completely irrelevant to the question whether A_0 is probabilistically justified or not.

Rather than talk about a nebula, we could also use the metaphor of a borehole. Compare the justification of a target by an epistemic chain to the pumping up of water from a deep well. If the chain is non-probabilistic, then the relations of entailment serve as neutral conduits through which justification passes unhindered. The justification itself comes from the bottom of the borehole, whence it is pumped up and transferred along the chain, whither it streams to the target proposition. If the epistemic chain is infinite, there is no beginning, the borehole is bottomless, the pumping stations forever remain dry, and no justification will ever gush out to the target. But now imagine that the infinite chain is probabilistic. Then a bottom is not needed. For now justification does not surge up unchanged from source to target; rather it comes from the conditional probabilities, which jointly work to confer upon the target proposition an acceptable probability. The conditional probabilities are, as it were, the intermediate pumping stations which actively take a moeity of justification from the circumambient earth, rather than passively wait for what comes up through the borehole. In a probabilistic regress we deliver justification, albeit piecemeal, whereas in a non-probabilistic regress we are not able to produce anything at all. In the latter case there is nothing more

---

[16] Pollock 1974, 26-31.

than the pointing to a fathomless borehole, or to a bank teller beyond the end of the universe who is supposed to administer my fortune.

Yet another metaphor was suggested to us by an anonymous reviewer; it concerns the saga of the bucket brigade. Suppose there is a fire and Abby gets her water from Boris, and Boris gets it from Chris, and Chris from Dan, and so on *ad infinitum*. It would seem that the fire will never be put out, since there is no first member of the brigade who actually dips his or her bucket into the lake. However, once we assume that justification involves probabilistic support the dousing operation looks quite different. Under this assumption, the proposition 'Abby gets water from Boris' (A_0) is only probabilistically justified, and we can calculate the probability value of A_0 by applying the rule of total probability that we cited earlier:

$$P(A_0) = P(A_0|A_1)P(A_1) + P(A_0|\neg A_1)P(\neg A_1), \tag{6.3}$$

where A_1 reads 'Boris gets water from Chris'. Of course, whether Boris gets water is also merely probable, and its probability depends on whether Chris gets water, and so on. We face here an infinite series of probability values calculated via the rule of total probability. As we know by now, we are perfectly able to compute the outcome of this infinite series in a finite time: with the numbers that we used in the uniform case of the bacterium example in Section 3.7, the probability that Abby gets water is $\frac{2}{3}$.

All four probabilities on the right-hand side of (6.3), the conditional as well as the unconditional ones, are supposed to have values strictly between zero and one (in the interesting cases). In contrast, the regress of entailments, in which justification is not probabilistic, can be modelled by restricting all four 'probabilities' to be 0 or 1. Within this non-probabilistic approach, Abby either gets water or she does not. According to the no starting point objection, the moral of the saga about the bucket brigade is precisely that she does not get water — if the number of brigadiers is infinite. Because this is unacceptable, it is concluded that there must be a first firefighter on the shore of the lake who starts off the whole operation. In the probabilistic scenario the existence of a primordial firefighter is not needed, since the problem that it is supposed to solve does not arise in the first place. The reason is, as we have seen, that now the relations between the propositions are not idle channels, but actively contribute to the probability value of A_0; they for example allow for a downpour somewhere along the line that fills the bucket. So if we take seriously that justification involves probabilistic support, then the probability that Abby extinguishes the fire can have a precise value, despite the infinite number of her team-mates. As in the examples that we considered above, this unconditional value is a function of all the conditional probabilities.

6.2 A Probabilistic Regress Needs No Starting Point

Note that the above reasoning is independent of whether we embrace an objective interpretation of probability (assuming, for example, that the firefighters have propensities for handing over the water only now and then) or a subjective interpretation (in which we specify our degree of belief in A_0). Both the objective and the subjective interpretation are bound by the rule of total probability, and that is all that counts here. This suggests that our approach is not restricted to epistemological series, but might be applied more generally to the metaphysical structures that Carl Gillet has been talking about. In fact, it might even be used to query similar reasonings in ethics. Richard Fumerton argued that his conceptual regress argument for foundationalism has a counterpart in the ethical realm. Suppose we are interested in whether an action, X, is good, and suppose we are being offered a series of conditional claims: if Y is good then X is good, if Z is good then Y is good, and so on, *ad infinitum*. Have we answered the original question? Fumerton believes we have not. At best we possess an infinite number of conditional claims, but this does not tell us whether X is good. Just as inferential justification only makes sense if there exists noninferential justification, instrumental goodness only makes sense if we assume that some things are intrinsically good:

> ...the view that there is only instrumental goodness is literally unintelligible. To think that something X is good if all goodness is instrumental is that X leads to a Y that is good by virtue of leading to a Z that is good, by virtue of ..., and so on ad infinitum. But this is a vicious conceptual regress. The thought that X is good, on the view that all goodness is instrumental, is a thought that one could not in principle complete. The thought that a belief is justified, on the view that all justification is inferential, is similarly, the foundationalist might argue, a thought that one could never complete.
>
> Just as one terminates a conceptual regress involving goodness with the concept of something being intrinsically good, so one terminates a conceptual regress involving justification with the concept of a noninferentially justified belief.[17]

> The concept of intrinsic goodness stands to the concept of instrumental goodness as the concept of noninferential justification stands to the concept of inferential justification. Just as there are no good things without there being something that is intrinsically good, so also there are no inferentialy justified beliefs unless there are noninferentially justified beliefs.[18]

Fumerton would be right that instrumental goodness implies intrinsic goodness if the conditional claims are of the form 'if Y is good then X is good'.

---

[17] Fumerton 1995, 90.
[18] Fumerton 2014, 76.

For then goodness is transferred lock, stock and barrel along the chain, and the no starting point objection, or rather Gillet's more general Structural Objection, applies in full force. However, we have been arguing that the situation changes radically if the claims take on the form 'if Y is good then there is a certain probability that X is good' and 'if Y is bad then there is a certain (lower) probability that X is good', and so on. For now goodness is not transferred in its entirety along the series. Rather it slowly emerges as we progress from the links Z to Y and Y to X. In this probabilistic scenario the original question would be how probable it is that a certain action, X, is good. And this question can indeed be answered; as we have seen, with the numbers chosen, it is $\frac{2}{3}$.

6.3 The Reductio Argument

According to the reductio argument, if an infinite chain could justify a target A_0, then another infinite chain could be constructed that would justify the target's negation, $\neg A_0$. Since it does not make sense for a proposition and its negation both to be justified, the proponents of this argument conclude that justification by an infinite chain is absurd.

Like the no starting point objection, the reductio argument has taken on different formulations. Here we will concentrate on a version that was offered by John Post in a tightly argued paper, which is in fact an improved version of arguments that had been put forward by John Pollock and James Cornman.[19]

Post starts his argument by defining an infinite justificational regress as a "non-circular, justification-saturated regress", by which he means that "every statement in the regress is justified by an earlier statement, and none is justified by any set of later statements".[20] As we have seen in Chapter 2, Post sees the justification relation as entailment, or better, 'proper entailment': "if anything counts as an inferential justification relation, proper entailment does ... If A_n properly entails A_{n-1}, then A_{n-1} is justified".[21] Now consider again the infinite chain

$$A_0 \longleftarrow A_1 \longleftarrow A_2 \longleftarrow A_3 \longleftarrow A_4 \longleftarrow \ldots \tag{6.4}$$

---

[19] Post 1980; Pollock 1974, 28-29; Cornman 1977.
[20] Post 1980, 3.
[21] Ibid. Post has X and Y where we write A_n and A_{n-1}.

6.3 The Reductio Argument

where it is assumed that the propositions are connected by proper entailment relations in the sense of Post, and where again A_0 does duty for the target q. According to Post, chain (6.4) is a non-circular, justification-saturated regress if and only if the following three conditions are satisfied:

a. A_n entails A_{n-1} $(n > 0)$;
b. A_n is not entailed by any $A_{m<n}$;
c. A_n is not justified on the basis of any set of $A_{m<n}$.

The first condition captures the idea that justification is a relation of entailment. The second condition is meant to ensure non-circularity. The third condition is added in order to block the possibility that a set of propositions might in some way or other together conspire to justify a proposition higher in the chain, which would make the regress circular after all. In the following we will always assume non-circularity in the background.

The construction of (6.4) as a non-circular, justification-saturated regress presupposes that at every step of the regress there indeed exists some proposition, A_n, which satisfies conditions a, b and c. Are there any examples of (6.4) that do the job? According to Post there are many, since there are many forms of proper entailment which meet the three conditions above. One of them is obtained by using *modus ponens* to interpret the links in the chain as follows:

$$A_0 = B_0$$
$$A_1 = B_1 \wedge (B_1 \rightarrow B_0)$$
$$A_2 = B_2 \wedge (B_2 \rightarrow (B_1 \wedge (B_1 \rightarrow B_0)))$$
$$A_3 = B_3 \wedge (B_3 \rightarrow (B_2 \wedge (B_2 \rightarrow (B_1 \wedge (B_1 \rightarrow B_0))))), \quad (6.5)$$

and by adding the restriction that B_1 is some proposition not entailed by A_0, that B_2 is some proposition not entailed by A_1, and so on. Under these restrictions it is the case that A_1 entails A_0, A_2 entails A_1, and so on; but A_0 does not entail A_1, A_1 does not entail A_2, and so on. Moreover, there is no set of propositions that together justify a proposition higher in the chain, so the conditions a, b and c are fulfilled.

Since $B \wedge (B \rightarrow A)$ is formally equivalent to $B \wedge A$, (6.5) can also be written as

$$A_0 = B_0$$
$$A_1 = B_1 \wedge B_0$$
$$A_2 = B_2 \wedge B_1 \wedge B_0$$
$$A_3 = B_3 \wedge B_2 \wedge B_1 \wedge B_0, \quad (6.6)$$

and so on, so that the chain (6.4) amounts to

$$B_0 \leftarrow (B_1 \wedge B_0) \leftarrow (B_2 \wedge B_1 \wedge B_0) \leftarrow (B_3 \wedge B_2 \wedge B_1 \wedge B_0) \leftarrow \ldots \quad (6.7)$$

Each link in (6.7) justifies its neighbour to the left, with the exception of B_0, which has no left-hand neighbour.[22]

Does it make sense to say that (6.7) justifies A_0? Post rightly claims that it does not. For in this manner a regress of propositions can be constructed for *any* target proposition, in particular for the negation of A_0. We only need to construct the infinite chain:

$$A'_0 \leftarrow A'_1 \leftarrow A'_2 \leftarrow A'_3 \leftarrow \ldots \quad (6.8)$$

where the A'_n are interpreted as

$$A'_0 = \neg B_0$$
$$A'_1 = B'_1 \wedge (B'_1 \rightarrow \neg B_0)$$
$$A'_2 = B'_2 \wedge (B'_2 \rightarrow (B'_1 \wedge (B'_1 \rightarrow \neg B_0)))$$
$$A'_3 = B'_3 \wedge (B'_3 \rightarrow (B'_2 \wedge (B'_2 \rightarrow (B'_1 \wedge (B'_1 \rightarrow \neg B_0))))). \quad (6.9)$$

Chain (6.8) reduces to

$$\neg B_0 \leftarrow (B'_1 \wedge \neg B_0) \leftarrow (B'_2 \wedge B'_1 \wedge \neg B_0) \leftarrow (B'_3 \wedge B'_2 \wedge B'_1 \wedge \neg B_0) \leftarrow \ldots \quad (6.10)$$

So if an infinite regress could justify a target proposition A_0, then another could justify $\neg A_0$, which is of course absurd. Hence the reductio argument, which shows that an infinite regress of proper entailments cannot justify a proposition.

Both Peter Klein and Scott Aikin made an attempt to ward off the reductio. Klein's idea is that an infinite chain of proper entailments as set up by Post is necessary, but not sufficient for the justification of a target: in order to be sufficient, the propositions in the chain should also be "available" as reasons.[23] Aikin has argued that the only way to repel the reductio argument is by taking a mixed view: infinitism and foundationalism do not exclude one another, for a proposition can be both inferentially and noninferentially

---

[22] Eq.(6.6) is used in Oakley's second argument against justification by infinite regress (Oakley 1976, 227-228). Aikin calls (6.6) "the simplification reductio" (Aikin 2011, 58.)

[23] Klein 1999, 312; Klein 2003, 722.

justified.²⁴ Aikin here takes up an idea by Jay Harker, namely that not all regresses of entailment make sense as justificatory chains, but that some do. According to Harker, a regress merely of beliefs is insufficient; a justificatory chain must contain relations to facts as well, although it may still be infinite.²⁵

Thus Klein, Aikin and Harker all endorse the intuition that more is needed for justification than an infinite, unanchored chain of proper entailments; something has to be added to this chain in order to make it a *justificatory* chain. We fully share this intuition, but we think that a chain of entailments does not lend itself so easily to such an add-on — it is somehow too self-contained for that. What helps to prevent the reductio is to abandon the idea that the links in the chain are connected via proper entailment and to adopt connections through probabilistic support. Holding on to the assumption of entailment means strenghtening the reductio argument; the argument is better combated by assuming regresses to be probabilistic, as we will explain in the following section.

6.4 How the Probabilistic Regress Avoids the Reductio

In a standard finite chain such as (6.1), where the arrow represents entailment, the ground A_{m+1} is all-important: the truth value of the target A_0 is a function of the truth value of A_{m+1} and of nothing else. The story is basically the same in the infinite case. However, there is then no ground, which is precisely the reason why it does not make sense to say that the target is justified. The concept of entailment is the culprit here, for it forces us to accept two things that are hard to combine, namely that the ground is all-important and non-existent at the same time. Exactly this combination precipitates the reductio argument. Nothing now restricts us in gratuituously constructing a rivalling regress that 'justifies' the target's negation, since the only restriction that matters, to wit the truth value of the ground, is conspicuous by its very absence.

The situation is entirely different in an infinite probabilistic regress. True, there too a ground is lacking. But this is irrelevant, for the probability value of the target is a function of the conditional probabilities alone. So it all de-

---

²⁴ Aikin 2011, 59-60 and Chapter 3.
²⁵ Harker 1984. Selim Berker takes a comparable route, offering the infinitist a way to avoid a fundamentalist regression stopper without running the gauntlet of the reductio argument — in Section 8.6 we will briefly come back to Berker.

pends on the question: What, in a justificatory chain, determines the value of the target? In a standard chain of entailments, the truth value of the target is determined by that of the ground, independently of the length of the chain. In a probabilistic chain, however, the length of the chain is relevant. If the probabilistic chain is finite, then the target's probability value is a function of both the unconditional probability of the ground and the conditional probabilities. As the chain gets longer, the influence of the ground decreases while the influence of the combined conditional probabilities increases. In the limit that the chain goes to infinity, only the conditional probabilities matter, and the rôle of the ground has died out (in the usual class). In this regard the difference between a non-probabilistic and a probabilistic regress could not be greater: in the former, the only variable that counts is a function of the ground, whereas in the latter the ground is of no significance whatsoever.[26]

We may conclude that the reductio argument misfires when the regress is a probabilistic one. The argument hinges on the assumption that the only variable which is responsible for the truth value of the target, namely the truth value of the ground, is non-existent. This absence of a ground allows us to concoct as many free-floating regresses as we wish, since the only variable that would determine the truth value of the target, viz. the truth value of the ground, is forever postponed and never actualized. In a probabilistic regress, on the other hand, the non-existent ground is not pertinent to the probability value of the target.

However, one could argue that this is too easy. For is it not possible to construct a rivalling probabilistic regress, i.e. a regress that supports the negation of our target? The only thing we would have to do is to come up with a set of conditional probabilities that numerically, and thus purely formally, bestow upon the target a probability value that for example exceeds the chosen threshold. If these conditional probabilities are not in any way connected to the world, we can cook them up *ad libitum*. We could then well end up with two rivalling probabilistic regresses, one probabilistically justifying A_0, and the other one probabilistically justifying $\neg A_0$.

Although the above argument is formally valid, it is not applicable to the issue that we are talking about. For it only works if the conditional probabilities are regarded as free variables, whose values may be chosen at will. We are however interested in epistemic justification, i.e. in the justification of propositions about our knowledge of how the world actually is, and this means that the conditional probabilities are not freely chosen. On the contrary, as we explained in Section 4.4, in a probabilistic regress the conditional

---
[26] See also Peijnenburg and Atkinson 2014a.

6.4 How the Probabilistic Regress Avoids the Reductio

probabilities carry all the empirical thrust. Once we admit empirically determined conditional probabilities, we are not free to invent other conditional probabilities in a competing regress for the negation of our target proposition: the conditional probabilities are determined too, and they yield a probability for the negation of the target that is one minus the probability of the target. If the target probability clears a threshold of acceptance greater than one half, the probability of the negation of the target will not do so.

Our opponent might not be satisfied, and complain that it remains unclear *how* conditional probabilities can carry empirical information; after all, the interface between our propositions and the world is fraught with difficulty. To this we would reply that, of course, such difficulties exist, and they are well documented; the problem of finding a transducer between our propositions and the world cuts deep and might even turn out to be insoluble. But as we made clear in Section 4.4, our aim is not to say something about that problem: we are not trying to formulate an answer to the sceptic. Rather our aim is to draw attention to probabilistic regresses and to phenomena such as those of fading foundations and of the emergence of justification, and to point out that these phenomena have consequences for the age-old objections to infinite regresses.

Andrew Cling has argued that an infinite regress can only justify a proposition if a certain condition is satisfied, notably that the regress is not "pure fiction" but has "grounding in how things are, are likely to be, or are reasonably believed to be".[27] The trouble with infinitism, says Cling, is that this condition can only be satisfied if simultaneously the very idea of justification by an infinite regress is undermined. Our analysis indicates that Cling is correct if the justificatory regress is a regress of entailments, not if it is probabilistic. For a probabilistic regress, as we have seen, can probabilistically justify a proposition while still having entry points for the world in the form of the conditional probabilities.

We have provisionally argued that these conditional probabilities arise from experiments, but of course they are not indubitable, and they can be questioned in turn. In that case they become the targets of new probabilistic chains. As we will explain in Section 8.5, this takes us from one-dimensional chains to multi-dimensional networks, where the effect of fading foundations still obtains.[28]

---

[27] Cling 2004, 111; see also Moser 1985, who makes a point similar to that of Cling.
[28] William Roche doubts whether a probabilistic regress can take away Cling's worry (Roche 2016). We think that it can indeed, for the reasons explained here and in Sections 4.4 and 8.5.

6.5 Threshold and Closure Constraints

It would be foolhardy to claim that probabilistic support along a chain of propositions or beliefs is sufficient for their justification. An obvious objection to such a claim would be that, after all the contributions from the conditional probabilities have been summed, the resulting probability of the target might turn out to be less than a half, which means that, relative to this particular chain, the target would be more likely false than true. Under these circumstances one would not say that the chain justifies the target. Indeed, as we have stressed, something must be added to probabilistic support to achieve a sufficient condition for justification.

Although it is certainly not our ambition to answer the difficult question of sufficiency, we shall in this section discuss two additional candidate desiderata for justification. The first is simply a threshold constraint on the target probability; the second is a modified threshold requirement for a measure of justification that has been proposed by Tomoji Shogenji. We first look at the simple threshold constraint, using the tables in Chapter 4 as illustration. We recall the well-known fact that this constraint falls foul of the intuition that justification should be closed under conjunction. But should unrestricted closure be a desideratum for justification? We argue that it should not: closure should be required only for *independent* propositions. The simple threshold constraint does not respect this modified closure requirement, and so it should be rejected. Shogenji's threshold condition, however, does respect this modified closure requirement. What makes Shogenji's condition especially interesting for us, moreover, is that it sails between entailment and probabilistic support: it is stronger than mere probabilistic support, but weaker than entailment. It is therefore a refined desideratum for justification; but we are not so incautious as to claim that it is a sufficient condition.

The simple threshold constraint amounts to the introduction of a context-dependent threshold of acceptance, say t, that is greater than one-half, but less than one.[29] As a first attempt, we might propose that if q is justified to degree t by a single proposition, or by a finite or infinite chain of propositions, then there must be probabilistic support along the chain, *and* $P(q)$ must be not less than t. Here is an example. Suppose that we take $t = \frac{3}{4}$ and refer to the tables in Chapter 4. We see from Table 4.1 that $P(q)$ does *not* clear $\frac{3}{4}$ with a chain of ten or fewer intermediate A's, but that it does so with a chain of twenty-five or more intermediate A's.

---
[29] Carnap 1980, 43, 70, 107; Fitelson 2013.

6.5 Threshold and Closure Constraints

For a second example, look at Table 4.2, and again let $t = \frac{3}{4}$. Now we see that $P(q)$ clears the threshold in all cases, even when there is only one intermediate A. The reason for this is simply that the ground p has a high probability; and in connection with the chosen values of the conditional probabilities α and β (0.99 and 0.04) this means that $P(q)$ already exceeds the threshold of $\frac{3}{4}$ after one step. Had α and β both been small, then the situation would have been very different; for then no number of steps would have been enough to reach the threshold, no matter how large the probability of p was. It can also happen that the value of $P(q)$ is larger than the threshold after a few steps, but sinks below the threshold if the chain gets longer. This can be illustrated by appealing to Table 4.2 again, and adopting the more demanding threshold of $t = 0.85$ instead of 0.75. With ten or fewer steps this more stringent threshold is exceeded, but with twenty-five or more steps we see that $P(q)$ has sunk below the new threshold. In such a case q might appear to be justified (to degree 0.85), but later, as the chain lengthens, we discover that this is not so.

Now consider still another example. Let the conditional probabilities both be very large, for example 0.99 and 0.96. Here again the target proposition, q, will have a probability well in excess of the threshold of $\frac{3}{4}$, even when there is only one intermediate A. And this is so irrespective of what the probability of p might be. Here the joint conditional probabilities are already doing all the work. On the other hand, if both conditional probabilities are very small, then the probability of q will be very small, again irrespective of $P(p)$. This is because the rule of total probability shows that $P(q)$ is an interpolation between the two conditional probabilities, $P(q|A_1)$ and $P(q|\neg A_1)$. In such a case the target could not be justified by the regress.

What these examples show is that the conditional probabilities, together with the unconditional probability of p, determine how long it takes before $P(q)$ reaches the threshold, if indeed it does so. Sometimes the unconditional probability of p has considerable influence, sometimes its influence is smaller: it is all contingent on the particular values. In the case of an infinite regress in the usual class, if the probability clears the threshold, this is achieved by the infinite set of conditional probabilities alone, without any contribution from p.

However, requiring that justification implies that the target probability meet a threshold of acceptance runs into difficulties, as we have intimated. For if target propositions q and q' are each supported by A_1, and if each meets some threshold, t, which is strictly less than one, it does not follow that the probability of the conjunction of q and q' meets t. Should we require that, if propositions q and q' are each separately *justified* by the same evidence A_1, then the proposition 'q and q'' is justified by the same evidence A_1? That is, should we require that justification is closed under conjunction? To see that an unqualified 'yes' would be too quick an answer, let us look at a simple example. Suppose that a fair die is tossed, but not yet inspected. Let q be the proposition 'the die shows 5', and q' be the proposition 'the die shows 6'. Let A_1 be the proposition 'the die shows more than 4'. Then $P(q) = P(q') = \frac{1}{6}$, and $P(q|A_1) = P(q'|A_1) = \frac{1}{2}$, so both q and q' are probabilistically supported by A_1. However, q and q' are incompatible with one another, so $P(q \wedge q') = 0$; and of course we would not want to claim that A_1 justifies the impossibility $q \wedge q'$. The conclusion is that we should not allow unlimited closure of justification under conjunction. This is of course the lesson that many people have drawn from the lottery paradox and similar quandaries concerning unrestricted closure of justification under conjunction. If one is justified in believing that ticket t_i in a fair lottery will lose, and that ticket t_j will also lose, is one justified in believing to the same extent that both t_i and t_j will lose? Evidently not, for the two failures to win are not independent of one another: if t_i loses, the chance that t_j will lose is reduced.

If unrestricted closure is forbidden, what would be a reasonable requirement concerning closure? Look at another example: suppose now that two coloured dice are tossed, but not yet inspected. Let q be the proposition 'the red die shows 5', and let q' be the proposition 'the blue die shows 6', and let A_1 be the proposition 'each die shows more than 4'. Once more $P(q) = P(q') = \frac{1}{6}$, and $P(q|A_1) = P(q'|A_1) = \frac{1}{2}$, so again both q and q' are probabilistically supported by A_1 to the same degree. Now q and q' are compatible, moreover they are independent of one another, both unconditionally and conditionally:

$$P(q \wedge q') = P(q)P(q') = \frac{1}{36}$$
$$P(q \wedge q'|A_1) = P(q|A_1)P(q'|A_1) = \frac{1}{4}.$$

Again A_1 supports q and q' probabilistically, but it also supports the conjunction, $q \wedge q'$, for $P(q \wedge q'|A_1) \geq P(q \wedge q')$. Note that the degree of probabilistic support that A_1 gives to the conjunction $q \wedge q'$ is not the same as the degree of support it gives to the conjuncts. However, if A_1 *justifies* $q \wedge q'$, then it *is* reasonable to require that A_1 justifies the conjunction to the same degree as it justifies the conjuncts. After all, if one is justified (to some extent) in expecting the red die to show 5, and also in expecting the blue die to show 6, on the basis of knowledge that each of the dice shows either 5 or 6, then one should be justified, to the same extent, in expecting that the red die shows 5 and the blue die shows 6, on the same knowledge basis. That the red die shows 5

6.5 Threshold and Closure Constraints

does not influence whether the blue die shows 6. Evidently the requirement that the probability clear a threshold of acceptance is not an adequate criterion; and it must be rejected as a desideratum for justification. The problem now is to find a measure of justification that clears a threshold *and* respects the findings of the above dice scenarios, and others like it.

Tomoji Shogenji has constructed just such a measure of justification.[30] Suppose that q and q' are independent, both unconditionally and also when conditioned by A_1. Suppose further that both q and q' have measures of justification greater than some threshold of acceptance, s. Then Shogenji requires that their conjunction $q \wedge q'$ also has a measure of justification greater than s. Thus his measure $J(q,A_1)$, the justification that A_1 bestows on q, respects closure in the restricted sense.

Measure $J(q,A_1)$ is a function of the various probabilities associated with q and A_1. But which function should it be? There are three independent candidates for the arguments of the function, for example $P(q)$, $P(A_1)$ and $\alpha_0 = P(q|A_1)$. Shogenji's first step is to strike out $P(A_1)$, on the grounds that, if one were to conjoin to A_1 some independent and irrelevant proposition, I, the justification that $A_1 \wedge I$ gives to q should be the same as that given by A_1. But $P(A_1 \wedge I) = P(A_1)P(I)$, and so the degree of justification *would* be changed by the conjunction if the measure were to depend on $P(A_1)$. So the required measure of justification must be a function, f, of $P(q)$ and $P(q|A_1)$ alone:[31]

$$J(q,A_1) = f[P(q), P(q|A_1)].$$

This immediately rules out the confirmation measure

$$S(q,A_1) = P(q|A_1) - P(q|\neg A_1),$$

as a candidate for a measure of justification, since that may be rewritten as

$$S(q,A_1) = \frac{P(q|A_1) - P(q)}{1 - P(A_1)},$$

which is manifestly a function of $P(A_1)$, as well as $P(q)$ and $P(q|A_1)$.[32]

Evidently the standard measure of confirmation, D,

$$D(q,A_1) = P(q|A_1) - P(q),$$

---
[30] Shogenji 2012.
[31] Note that $P(q|A_1 \wedge I) = P(q|A_1)$, if I is independent of A_1 and of $q \wedge A_1$.
[32] $S(q,A_1)$ is of course the same as γ_0.

does satisfy Shogenji's first desideratum for J. As we remarked in Chapter 2, Carnap called this an "increase in firmness", the extent to which the probability of q is increased by conditioning it on A_1. Shogenji requires that $J(q,A_1)$ should increase if $P(q|A_1)$ increases while $P(q)$ is held fixed, and decrease if $P(q)$ increases while $P(q|A_1)$ is held fixed. It is clear that the measure D does these things.

Could D be the required measure of justification, J? Not so, as we can see from the example of the coloured dice, since

$$D(q,A_1) = D(q',A_1) = \tfrac{1}{2} - \tfrac{1}{6} = \tfrac{1}{3}$$
$$D(q \wedge q', A_1) = \tfrac{1}{4} - \tfrac{1}{36} = \tfrac{2}{9},$$

which are different, whereas the degree of justification of the conjunction of the independent propositions q and q' should be the same as that for q and q' separately. But not only does D not satisfy this closure requirement, none of the many other measures of confirmation do so either![33]

Shogenji shows that the following new measure does satisfy the requirement of closure:

$$J(q,A_1) = 1 - \frac{\log P(q|A_1)}{\log P(q)}. \tag{6.11}$$

Although this is not the only function that satisfies Shogenji's desiderata for a measure of justification, it has been proved that all functions that do so are ordinally equivalent to Shogenji's J function.[34] That is to say, if A_1 gives a higher degree of justification to one proposition than it does to another, according to the measure (6.11), then this ordering of justificatory degrees will be the same for any other measure that satisfies Shogenji's conditions. We may say that the measure (6.11) is the unique solution of the problem, up to ordinal equivalence. A proof of the above is given in Appendix B; but here we shall simply check that the Shogenji measure works properly for our coloured dice. From (6.11) we calculate

$$J(q,A_1) = J(q',A_1) = 1 - \frac{\log \tfrac{1}{2}}{\log \tfrac{1}{6}}$$

$$J(q \wedge q', A_1) = 1 - \frac{\log \tfrac{1}{4}}{\log \tfrac{1}{36}} = 1 - \frac{2\log \tfrac{1}{2}}{2\log \tfrac{1}{6}} = 1 - \frac{\log \tfrac{1}{2}}{\log \tfrac{1}{6}}.$$

---

[33] See Atkinson, Peijnenburg and Kuipers 2009 for a list of ten measures of confirmation. A seminal paper on different measures of confirmation is Fitelson 1999.
[34] Atkinson 2012.

Thus $J(q,A_1) = J(q',A_1) = J(q \wedge q', A_1)$, so if $J(q,A_1) \geq s$ and $J(q',A_1) \geq s$, for some s, it is trivially the case that $J(q \wedge q', A_1) \geq s$. In words, if q and q' are Shogenji-justified to the same degree, their conjunction is also Shogenji-justified to that degree, as should be the case.

If the degree of Shogenji justification that A_1 gives to q is not less than s, i.e. $J(q,A_1) \geq s$, then
$$1 - \frac{\log P(q|A_1)}{\log P(q)} \geq s,$$
and this can be recast in the form (see Appendix B)
$$P(q|A_1) \geq [P(q)]^{1-s}. \tag{6.12}$$

Note that when $s = 0$ — so there is effectively no threshold — this inequality reduces to
$$P(q|A_1) \geq P(q)$$
which is equivalent to our condition of probabilistic support (or neutrality, in the case of the equals sign). On the other hand, when the threshold is at its maximum, so that $s = 1$, the relation becomes
$$P(q|A_1) \geq 1, \quad \text{which of course implies} \quad P(q|A_1) = 1,$$
since no probability can be greater than one. This is the probabilistic condition that corresponds to entailment.

For non-extremal values of the degree s, the measure J interpolates between probabilistic support and entailment. Since entailment is too strong a requirement for a viable understanding of justification, and probabilistic support is too weak, it is very suggestive that this measure of Shogenji may be a step in the right direction in the search for the holy grail of a sufficient condition for justification.

6.6 Symmetry and Nontransitivity

In this chapter we have discussed the two conceptual objections to infinite epistemic chains that occur most frequently in the literature, the no starting point objection and the reductio argument, and we argued that they lose their bite when justification is seen as something that involves probabilistic support rather than entailment. Since probabilistic support is not enough for justification, we looked in the previous section at two candidates for add-ons.

One could however raise objections to the very concept of probabilistic support itself. It is after all the child of a theory that is beset by a number of serious pitfalls: the problem of old evidence, the problem of spurious relations, of irrelevant conjunctions, of randomness, and more.

Whenever a theory encounters problems, either we reject it because the problems are too serious, or we continue to use it, trying in the meantime to put things right. In the case of Kolmogorovian probability theory the choice seems clear. Aside from exotics such as quantum probability and Robinsonian nonstandard analysis, Kolmogorov's calculus is very much the only game in probability town. When in epistemology we say that one proposition 'probabilifies' another, it would be wise to take Kolmogorov's system seriously, at least until we have found a better interpretation of 'probabilifies'.

This book is not the place to dwell on all the snags and hitches of Kolmogorovian probability. Yet there are two properties of the concept of probabilistic support that require some further consideration, since epistemologists may find them troublesome in the context of epistemic justification. The first is the fact that probabilistic support is *not transitive* and the second one is that it is *symmetric*.

Many epistemologists have explicitly or implicitly expressed the view that epistemic justification is transitive: if A_n is justified by A_{n+1} and A_{n+1} is justified by A_{n+2}, then A_n is justified by A_{n+2}. Such a view is of course apposite if justification is perceived as entailment or implication, for then justification is transmitted unchanged from one proposition to another. But if justification is understood as involving probabilistic support, then transitivity may be violated. It all depends on what must be added to the relation of probabilistic support to yield that of justification. For example, if justification were equivalent to probabilistic support plus the Markov condition, then justification would be transitive, since transitivity is a property of probabilistic support when the Markov restriction is in place. If however justification were equivalent to probabilistic support plus a threshold condition, then it would not be transitive. As we have made clear, we refrain from making any claims about what has to be added to probabilistic support in order to yield justification. The point to make here is just that probabilistic support as a necessary condition for justification entails nothing about the transitivity of justification.

A similar argument applies to the required asymmetry of justification. When considered qualitatively, probabilistic support is symmetrical: if A_{n+1} supports A_n, then A_n supports A_{n+1}. However, from the fact that probabilistic support is (qualitatively) symmetric, it does not follow that *justification* is qualitatively symmetric as well. An argument parallel to the one just given about transitivity shows that the symmetry of probabilistic support

6.6 Symmetry and Nontransitivity

entails nothing about the symmetry of justification. In fact the example of the Markov condition fits the bill here, too. For if A_{n+1} supports A_n, and A_{n+1} screens off A_n from all 'ancestor' propositions in the chain, i.e. A_m where $m > n+1$, then A_n will in general not screen off A_{n+1} from all 'descendent' propositions, i.e. A_m where $m < n$. Thus if justification were equivalent to probabilistic support plus the Markov condition, it would not be qualitatively symmetric. As we stressed above, the Markov model is not meant to be taken as a serious candidate as to how justification should be defined: it merely shows that justification can be asymmetric, even though probabilistic support is symmetric. A formal demonstration of this fact is as follows. Consider these three statements:

(1) if A_{n+1} justifies A_n, then A_{n+1} probabilistically supports A_n
(2) if A_{n+1} probabilistically supports A_n, then A_n probabilistically supports A_{n+1}
(3) if A_{n+1} justifies A_n, then A_n justifies A_{n+1}.

The point is that (3) does not follow from (1) and (2). What does follow from the latter two statements is:

(3′) if A_{n+1} justifies A_n, then A_{n+1} probabilistically supports A_n and A_n probabilistically supports A_{n+1}.

The consequent of (3′) expresses the fact that probabilistic support is symmetric. But this does not mean that justification is symmetric; it does not follow from this that A_n justifies A_{n+1}.

It is important to note that the matter is quite different with respect to fading foundations. The effect of fading foundations is not a property like transitivity or symmetry. As a result, it does follow that justification implies the existence of fading foundations (within the usual class). In detail:

(1″) if A_{n+1} justifies A_n, then A_{n+1} probabilistically supports A_n
(2″) if A_{n+1} probabilistically supports A_n, and the conditional probabilities belong to the usual class, then fading foundations ensue
(3″) if A_{n+1} justifies A_n, and the conditional probabilities belong to the usual class, then fading foundations ensue.

In this case (3″) does follow from (1″) and (2″). Irrespective of whether we are talking about probabilistic support or about epistemic justification, the phenomenon of fading foundations is the same, the reason being that the latter does not have a meaning independent of probability theory, which we take to be necessary for justification: if there is no probabilistic support, then there is no justification. The properties of transitivity and symmetry, on the other hand, do not need to refer to probability theory in order to have the meanings that they have. Thus under justification the influence of the probability of the ground on the probability of the target decreases as the number of links in the chain increases. And in the limit that the number of links goes to infinity, this probabilistic influence vanishes completely, leaving the probability of the target fully independent of the probability of the ground.

Open Access This chapter is licensed under the terms of the Creative Commons Attribution 4.0 International License (http://creativecommons.org/licenses/by/4.0/), which permits use, sharing, adaptation, distribution and reproduction in any medium or format, as long as you give appropriate credit to the original author(s) and the source, provide a link to the Creative Commons license and indicate if changes were made.

The images or other third party material in this chapter are included in the chapter's Creative Commons license, unless indicated otherwise in a credit line to the material. If material is not included in the chapter's Creative Commons license and your intended use is not permitted by statutory regulation or exceeds the permitted use, you will need to obtain permission directly from the copyright holder.

Chapter 7
Higher-Order Probabilities

Abstract
At first sight, a hierarchical regress formed by probability statements about probability statements appears to be different from the probabilistic regress of the previous chapters. After all, the former involves higher and higher-order probabilities, whereas the latter is an epistemic chain in which one proposition or belief probabilistically supports another. Closer examination, however, teaches us that the two regresses are in fact isomorphic. A model based on coin-making machines demonstrates that the hierarchical regress is consistent.

7.1 Two Probabilistic Regresses

We have extensively discussed chains of propositions which probabilistically support one another. But in Chapter 3 we did mention that Lewis, and independently Russell, seemed sometimes to be talking about higher-order probability statements rather than about straightforward chains of propositions.[1]

The ambiguity is understandable enough. As we have seen, both Lewis and Russell took the view that probability statements like 'q is probable' or 'the probability of q is x' only make sense if one assumes that something else is

---

[1] Section 3.2, footnote 8, and Section 3.3, footnote 22. *Cf.* Reichenbach 1952, 151, where mention is also made of a probability of a probability. Roderick Chisholm has taken issue with Reichenbach's idea (especially as it is expressed in Reichenbach 1938), but in turn received criticism from Bruce Aune (Chisholm 1966, 22 *ff*; Aune 1972).

certain. The question then arises what exactly this 'something else' could be, and two answers appear to be natural.

According to the first, the 'something else' is the *reference class* on the basis of which the unconditional probability of q is determined. Lewis, we recall, argued that 'the probability of q is x' is in fact elliptical for 'the probability of q is x, on condition that A_1'. In many cases A_1 will be assumed to be certain, and thus to have probability unity. If this assumption is not made, then one has to assume that A_2 is certain in 'the probability of A_1 is x, on condition that A_2'. This reasoning forms the background to Lewis's conclusion that a regress of probability statements only makes sense if it is rooted in a certainty. According to the second answer, however, it is the *entire probability statement* that is taken to be certain. In asserting 'the probability of q is x', one usually presupposes that this assertion itself has probability unity. If one does not, then one might assume that the probability that the probability of this assertion is y (with y smaller than one) is certain. In other words, with the abbreviation of 'the probability of q is x' as A_1, one way in which A_1 could fail to be certain is if the assertion 'the probability of the probability that A_1 is y' (call this assertion A_2) is one.

These two answers lead to two different readings of a probabilistic regress. According to the first, the regress states (with v_n standing for the unconditional probability values):

the probability of q, on condition that A_1 is true, is v_0;
the probability of A_1, on condition that A_2 is true, is v_1;
the probability of A_2, on condition that A_3 is true, is v_2;
and so on.

According to the second reading, the regress amounts to:

A_1: the probability of q is v_0;
A_2: the probability of A_1 is v_1;
A_3: the probability of A_2 is v_2;
and so on.

In the first kind of regress every A_n represents a condition on the probability of q or on that of A_{n-1}. In the presence of such a regress, as we have seen, we generally are able to determine the unconditional probability of q via an infinite iteration of the rule of total probability. In fact, as we have explained, the iteration need not even be infinite in order for us to compute the unconditional probability of q to an acceptable approximation. However, in the second regress every A_n names a statement about a probability. It thus

7.2 Second- and Higher-Order Probabilities

involves infinitely many statements about ever and ever higher-order probabilities, whereas the first regress refers to an infinite number of conditions.

Up to this point we have concentrated on the first kind of regress. In this chapter we shall focus on probabilistic regresses of the second kind, culminating in infinite series of probability statements about probability statements. We start in Section 7.2 by discussing probability statements of second and higher order. We will see that, although second-order probabilities do not pose any particular problem, many philosophers have objected to probability statements of a higher than second order. Especially the indefinite accumulation of probabilities to infinity has been generally regarded as not making sense.

In Section 7.3 we discuss an objection that Nicholas Rescher made to infinite-order probabilities. Our analysis of Rescher's argument will reveal that the above mentioned two readings of a probabilistic regress are in fact isomorphic, and in 7.4 this isomorphy will be demonstrated in a more formal way. Since regresses under the first reading are coherent, the isomorphy tells us that those under the second reading are too. Thus the properties of regresses under the first reading, such as those of fading foundations and emerging justification, are also properties of regresses under the second reading. In Section 7.5 we make the concept of infinite-order probability statements explicit by describing an executable model.

7.2 Second- and Higher-Order Probabilities

Suppose that the probability of the target proposition q is v_0:

$$P(q) = v_0. \tag{7.1}$$

If we know that (7.1) is true, then there is no more to be said; but what if we lack this knowledge? In that case, we may only be in a position to assert a further probabilistic statement like

$$P(P(q) = v_0) = v_1, \tag{7.2}$$

which is a second order probability statement, saying that the probability that (7.1) is true is v_1. Does (7.2) make sense? It can be argued that it does not. For if one supposes that (7.1) implies that $P(q) = v_0$ is true, then $P(P(q) = v_0) = 1$, and so, unless $v_1 = 1$, (7.2) would be inconsistent with (7.1). A way to avoid such an inconsistency would be to introduce two different probability functions instead of one, viz. $P^{(1)}$ and $P^{(2)}$. For evidently the intention is that (7.2) should *adjust* the initial bald statement (7.1). Thus we need to replace (7.1) and (7.2) by

$$P^{(1)}(q) = v_0$$
$$P^{(2)}(P^{(1)}(q) = v_0) = v_1, \qquad (7.3)$$

where $P^{(2)}$ is a second-order probability function.

However, objections have been raised against second-order functions like $P^{(2)}$, based on the contention that it is unclear what they mean. David Miller even argued that they lead to an absurdity.[2] In his view the only way second-order probability statements could make sense, if at all, would be if the second-order probability of q, given that the first-probability of q is v_0, is itself v_0:

$$P^{(2)}\left(q|P^{(1)}(q) = v_0\right) = v_0. \qquad (7.4)$$

He then goes on to argue that (7.4) leads to an unacceptable conclusion. For if we replace v_0 in (7.4) by $P^{(1)}(\neg q)$, we obtain

$$P^{(2)}\left(q|P^{(1)}(q) = P^{(1)}(\neg q)\right) = P^{(1)}(\neg q),$$

which is the same thing as

$$P^{(2)}\left(q|P^{(1)}(q) = \tfrac{1}{2})\right) = P^{(1)}(\neg q).$$

However, if instead we put $\tfrac{1}{2}$ for v_0 in (7.4), we find $P^{(2)}\left(q|P^{(1)}(q) = \tfrac{1}{2}\right) = \tfrac{1}{2}$. Therefore $P^{(1)}(\neg q) = \tfrac{1}{2}$, and thus $P^{(1)}(q) = \tfrac{1}{2}$. So if (7.4) were unrestrictedly valid, we could prove that the probability of an arbitrary proposition q is equal to one-half, which is absurd. This is known as the Miller paradox.

Brian Skyrms has argued against Miller's reasoning. Although Skyrms maintains that (7.4) is perfectly acceptable, playfully dubbing it 'Miller's Principle', he points out that Miller's further reasoning is fallacious, since it "rests on a simple *de re–de dicto* confusion".[3] As Skyrms explains, one and the same expression is used both referentially and attributively, so that a number (here v_0) is wrongly put on a par with a random variable, here $P^{(1)}(\neg q)$, that takes on a range of possible values.[4] So long as we recognize this confusion and keep the two levels apart, the notion of a second-order probability is harmless, and the Miller paradox disappears. We agree with

---
[2] Miller 1966.
[3] Skyrms 1980, 111.
[4] See Howson and Urbach 1993, 399-400, who make a similar observation.

7.2 Second- and Higher-Order Probabilities

Skyrms that Miller's Principle as such is harmless, but in what follows we will not need it: our reasoning goes through without the principle.

In addition to parrying Miller's argument, Skyrms warded off another objection to second-order probability statements, namely one that can be discerned in de Finetti's work. As is well known, de Finetti held that probability judgements are expressions of attitudes that lack truth values. Skyrms however pointed out that de Finetti's work is not particularly hostile to a theory of second-order probabilities:

> For a given person and time there must be, after all, a proposition to the effect that that person then has the degree of belief that he might evince by uttering a certain probability attribution.
>
> De Finetti grants as much:
>
> > The situation is different of course, if we are concerned not with the assertion itself but with whether 'someone holds or expresses such an opinion or acts according to it,' for this is a real event or proposition. (de Finetti 1972, 189)
>
> With this, de Finetti grants the existence of propositions on which a theory of higher-order personal probabilities can be built, but never follows up this possibility.[5]

De Finetti and Skyrms are not alone in having taken the view that second-order probabilities need not pose any particular problem. Several other authors recognize that, when the relevant distinctions are taken into account, second-order probabilities can be shown to be formally consistent.[6] This is not to say that such probabilities are mandatory. As Pearl has explained, second-order probabilities, although consistent, can be dispensed with, for one can always express them by using a richer first-order probability space.[7]

Once we accept the cogency of second-order probabilities, there is no impediment to constructing probabilities to any finite order.[8] We could continue the sequence (7.3) and introduce a hierarchy of higher-order probability statements:

$$P^{(1)}(q) = v_0$$
$$P^{(2)}(P^{(1)}(q) = v_0) = v_1$$
$$P^{(3)}(P^{(2)}(P^{(1)}(q) = v_0) = v_1) = v_2, \quad (7.5)$$

---

[5] Skyrms 1980, 113-114.
[6] Uchii 1973; Lewis 1980; Domotor 1981; Kyburg 1987; Gaifmann 1988.
[7] Pearl 2000.
[8] See Atkinson and Peijnenburg 2013.

and so on, with $P^{(m)}$ being the mth-order probability. In the previous section we introduced A_1, A_2 and A_3 as names of probability statements. Here we shall specify the probabilities in question more fully by stipulating their orders:

$$A_1 \text{ is the proposition } P^{(1)}(q) = v_0$$
$$A_2 \text{ is the proposition } P^{(2)}(A_1) = v_1$$
$$A_3 \text{ is the proposition } P^{(3)}(A_2) = v_2,$$

and so on. With these definitions, (7.5) can be written as

$$P^{(1)}(q) = v_0$$
$$P^{(2)}(A_1) = v_1$$
$$P^{(3)}(A_2) = v_2, \tag{7.6}$$

and so on.

However, the key question is of course not whether any finite series of higher-order probability statements is cogent, but whether the notion of infinite-order probabilities make sense. Is it coherent to continue the above sequence *ad infinitum*, in the limit defining a probability, $P^{(\infty)}(q)$, of infinite order? Leonard Savage has answered this question in the negative. For him, the mere fact that second-order probabilities provoke the introduction of probability statements of infinite order was enough to discard them altogether:

> Once second order probabilities are introduced, the introduction of an endless hierarchy seems inescapable. Such a hierarchy seems very difficult to interpret, and it seems at best to make the theory less realistic, not more.[9]

His conclusion is that "insurmountable difficulties" will arise if one opens the door to second-order probabilities and starts using such phrases as "the probability that B is more probable than C is greater than the probability that F is more probable than G".[10]

Savage was mainly talking about statistics, but in philosophy too it has been argued that an infinite order of probabilities of probabilities leads to problems that are insuperable. Thus David Hume argued in *A Treatise of Human Nature* that an infinite hierarchy implies that the probability of the target will always be zero:

---

[9] Savage 1954, 58.
[10] Ibid.

7.2 Second- and Higher-Order Probabilities 149

> Having thus found in every probability ...a new uncertainty ...and having adjusted these two together, we are oblig'd ...to add a new doubt.... This is a doubt ...of which ...we cannot avoid giving a decision. But this decision, ...being founded only on probability, must weaken still further our first evidence, and must itself be weaken'd by a fourth doubt of the same kind, and so on in infinitum: till at last there remain nothing of the original probability, however great we may suppose it to have been, and however small the diminution by every new uncertainty.[11]

Nicholas Rescher, in his book *Infinite Regress: The Theory and History of Varieties of Change*, also argued against an infinite hierarchy of probabilities. As he sees it, the problem with such a hierarchy is not that the probability of the target q will always be zero, but rather that it becomes impossible to know what that probability is:

> ...unless some claims are going to be categorically validated and not just adjudged probabilistically, the radically probabilistic epistemology envisioned here is going to be beyond the prospect of implementation. ...If you can indeed be certain of nothing, then how can you be sure of your probability assessments. If all you ever have is a nonterminatingly regressive claim of the format ...the probability is .9 that (the probability is .9 that (the probability of q is .9)) then in the face of such a regress, you would know effectively nothing about the condition of q. After all, without a categorically established factual basis of some sort, there is no way of assessing probabilities. But if these requisites themselves are never categorical but only probabilistic, then we are propelled into a vitiating regress of presuppositions.[12]

---

[11] Hume 1738/1961, Book I, Part IV, Section I. See also Lehrer 1981, for similar reasoning. As we noted in Section 3.3, Quine states in his lectures on Hume that this Humean argument is incorrect, since an infinite product of factors, all less than one, can be convergent, yielding a non-zero probability for the target (Quine 2008). Quine is right to point out this possibility, but note that it corresponds to what happens in our exceptional class, not in the usual class. Moreover, the possibility can only serve as a critique of Hume if one forgets about the second term in the rule of total probability, i.e. if all the $\beta_n = P(A_n | \neg A_{n+1})$ are zero. As we have seen, Hume does indeed leave out that term, as would Lewis and Russell many years later. Hume's argument is therefore not generally valid. Thus Quine's analysis of Hume is based on two unwarranted assumptions: first he assumes that all the conditional probabilities α_n belong to the exceptional class, i.e. the class of quasi-bi-implication, and second he supposes that all the conditional probabilities β_n are zero. What diminishes as the chain lengthens is not the probability of the target, as Hume and Quine thought, but rather the incremental changes that distant links bring about.

[12] Rescher 2010, 36-37. Rescher has p rather than q. Furthermore, Rescher explicitly conditions all his probabilities with respect to some evidence, E, and therefore

The argument of Rescher may seem plausible and persuasive. Yet we shall argue in the next section that an endless hierarchy of probabilities is in fact no stumbling block to having effective knowledge about the probability that q is true, let alone that it constitutes "an unsurmountable difficulty", as Savage would have it. In a sense the opposite is the case. If there is a stumbling block, it resides in the finite, not in the infinite hierarchy. For an infinite hierarchy of probabilities is, to a certain extent, better equipped to reveal the probability of q than is a finite one. The reason is reminiscent of the reason why a probabilistic regress of the sort that we have investigated in the previous chapters is cogent: in order to compute an infinite sequence, only the conditional probabilities need be known, whereas the computation of a finite sequence requires also knowledge of an unconditional probability.

7.3 Rescher's Argument

In this section we shall examine Rescher's claim: "If all you ever have is a nonterminatingly regressive claim of the format ... the probability is .9 that (the probability is .9 that (the probability of q is .9)) then in the face of such a regress, you would know effectively nothing about the condition of q", which amounts to putting v_0, v_1 and v_2 in (7.6) all equal to 0.9. We will show in this section that Rescher's assertion is in fact ill-founded.

Imagine, following Rescher, that we have a probability statement of the third order:
$$P^{(3)}(P^{(2)}(P^{(1)}(q) = 0.9) = 0.9) = 0.9. \tag{7.7}$$

Some philosophers conclude on the basis of (7.7) that the unconditional probability of q is 0.9, since no matter how many times one iterates, the probability value always stays the same.[13] This conclusion is also incorrect, but the question remains as to what *is* the correct conclusion that can be drawn from (7.7) about the unconditional probability of q.

Consider the definitions

A_1: The first-order probability $P^{(1)}$ of q is 0.9,
A_2: The second-order probability $P^{(2)}$ of A_1 is 0.9,
A_3: The third-order probability $P^{(3)}$ of A_2 is 0.9,

---

instead of Eqs.(7.1)–(7.2) he has $Pr(p|E) = v_0$ and $Pr(Pr(p|E) = v_0|E) = v_1$ (ibid., 36 — misprint corrected). In the interest of notational brevity, explicit reference to E will be suppressed.

[13] See for example DeWitt 1985, 128.

7.3 Rescher's Argument

and so on. In the rest of this section we will temporarily suppress the orders of the probabilities to facilitate an intuitive grasp of the course of the reasoning. So we have:

A_1: The probability of q is 0.9,
A_2: The probability of A_1 is 0.9,
A_3: The probability of A_2 is 0.9,

and so on. We will now successively revise $P(q)$; in the next section, when we approach the matter more formally, we will reinstate the higher orders.

We call on the rule of total probability,

$$P(q) = P(q|A_1)P(A_1) + P(q|\neg A_1)P(\neg A_1), \tag{7.8}$$

in which the probability of q is conditioned on that of A_1. In order to evaluate the unconditional probability of A_1, this formula must be repeated in the familiar way, with A_1 in the place of q, and A_2 in the place of A_1,

$$P(A_1) = P(A_1|A_2)P(A_2) + P(A_1|\neg A_2)P(\neg A_2), \tag{7.9}$$

and so on. Is it possible to calculate $P(q)$ if the format goes on to infinity? Rescher thinks not. If the hierarchy is endless one cannot know anything about the probability of q, for "we are propelled into a vitiating regress of presuppositions"[14]. The situation looks like a probabilistic analogue of the Tortoise's interminable query to Achilles, where the latter successively satisfies the former *pro tem* in higher and higher-order querulousness without end.[15]

However, this similarity is only apparent. Between the probabilistic and the nonprobabilistic version of the Tortoise's challenge to Achilles there is an essential difference: the latter might be hopeless, the former is not. It is true that the Tortoise can always ask about an unknown $P(A_n)$ after the weary warrior has taken n steps in his argument. It is also true that the unknown $P(A_n)$ could have any value between zero and one. However, the influence that $P(A_n)$ has on the value of $P(q)$ will be smaller as the distance between A_n and q gets bigger — even if $P(A_n)$ were to take on the largest allowed value of 1, see Section 4.3. As we know now, in the limit that n tends to infinity, the influence of $P(A_n)$ on $P(q)$ will peter out completely, leaving the value of $P(q)$ as a function of the conditional probabilities alone. Note again that this is not because $P(A_n)$ itself becomes smaller as n becomes larger: indeed,

---

[14] Rescher 2010, 37.
[15] Carroll 1895.

it may not do so. Nor is it simply because the iteration of (7.8) and (7.9), etc. leads to a series of terms that is convergent. Rather it is because $P(A_n)$ is multiplied by a factor that goes to zero as n tends to infinity. Each time Achilles has taken one more step, and the Tortoise has asked about $P(A_{n+1})$, this worrisome probability is multiplied by an even smaller factor, and after yet another step the Tortoise's $P(A_{n+2})$ is multiplied by a yet smaller factor still, and so on, until the factor has shrunk to zero.

Referring back to (7.8), we know from Miller's Principle that the term $P(q|A_1)$ is equal to 0.9. In Rescher's example, the third term, $P(q|\neg A_1)$, is not specified, but it will be clear that (7.8) cannot be evaluated without it: as long as the value of the third term is unknown, one cannot determine $P(q)$. For the sake of argument, we shall set this term equal to 0.3. It should be noted that no strings are attached to this choice of 0.3, since the argument is robust: whatever nonzero value of $P(q|\neg A_1)$ is chosen, so long as it is less than $P(q|A_1)$, the same reasoning will work.

Now (7.8) can be worked out:

$$P(q) = [0.9 \times 0.9] + [0.3 \times 0.1] = 0.84. \tag{7.10}$$

The number 0.84 was arrived at on the provisional assumption that the second term, $P(A_1)$, indeed equals 0.9, which would be correct if it were the case that

$$P(P(A_1) = 0.9) = P(A_2) = 1.$$

But that is wrong, for $P(A_2) = 0.9$. This means that $P(A_1)$ should rather be

$$P(A_1) = [0.9 \times 0.9] + [0.3 \times 0.1] = 0.84, \tag{7.11}$$

where, similarly, 0.3 is taken to be the value of $P(A_1|\neg A_2)$, and so on. On the basis of this new result, the value of $P(q)$ in (7.10) must be revised, yielding

$$P(q) = [0.9 \times 0.84] + [0.3 \times 0.16] = 0.804. \tag{7.12}$$

However, the number 0.804 was arrived at on the fictional assumption that the second term in (7.11), to wit $P(A_2)$, indeed equals 0.9, and thus that

$$P(P(A_2) = 0.9) = P(A_3) = 1.$$

But that is also wrong, for $P(A_3) = 0.9$. This means that $P(A_2)$ should rather be

$$P(A_2) = [0.9 \times 0.9] + [0.3 \times 0.1] = 0.84. \tag{7.13}$$

7.3 Rescher's Argument

On the basis of this, $P(A_1)$ is revised to

$$P(A_1) = [0.9 \times 0.9] + [0.3 \times 0.1] = 0.804. \tag{7.14}$$

This new value for $P(A_1)$ implies that $P(q)$ must again be revised, generating

$$P(q) = [0.9 \times 0.804] + [0.3 \times 0.196] = 0.7824, \tag{7.15}$$

and so on. It should be noted that these 'revisions' of the value of $P(q)$ are really higher and higher-order probabilities of q. We have suppressed the specification of the orders for greater readability: in the next section the technique will be explained with more care and with greater generality.

Here is an overview of the values that $P(q)$ takes after an increasing number of revisions:

Table 7.1 Unconditional probability of q after n revisions

n	1	2	3	5	10	15	20	∞
$P(q)$	0.84	0.804	0.7824	0.7617	0.7509	0.75007	0.750005	$\frac{3}{4}$

There are three important lessons to be drawn from these seemingly tedious calculations.

The first is that an endless hierarchy of probabilities can indeed determine what the probability of the original proposition is — contrary to what Rescher and many others have claimed. For it is possible to calculate the value of $P(q)$, even in a situation such as the one sketched by Rescher, where

$$P(P(P(q) = 0.9) = 0.9) = 0.9, \tag{7.16}$$

and so on. With the value that was chosen for $P(A_n|\neg A_{n+1})$, namely 0.3, and after an infinite number of revisions, $P(q)$ is exactly equal to $\frac{3}{4}$.

The second lesson is that an infinite number of revisions is not needed to come very close to the actual value of $P(q)$. For, as can be seen in Table 7.1, there is only a small difference between the value of $P(q)$ after, say, twenty revisions and after an infinite number of them. Of course, the size of the difference will depend on the numbers that are chosen for the conditional and unconditional probabilities in the equations: had the values of the first two terms been, for example, 0.8 rather than 0.9, and had $P(A_n|\neg A_{n+1})$ been 0.4 rather than 0.3, then not even twenty steps would have been needed to come as close to the limit value (which would have been $\frac{2}{3}$ in that case). There is always *some* finite number of revisions, such that the result scarcely differs from what is obtained with an infinite number of them.

The point can be regarded as a quantitative reinforcement of a claim that Rescher makes in qualitative terms. Partly in the wake of Kant and Peirce, Rescher stresses several times that some infinite regresses should be approached in a pragmatic way, in which it is acknowledged that contextual factors play an important role and that, at a certain point, "enough is enough":

> ...in any given context of deliberation the regress of reasons ultimately runs out into 'perfectly clear' considerations which are (contextually) so plain that there just is no point in going further. It is not that the regress of validation ends, but rather that we stop tracking it because in the circumstances there is no worthwhile benefit to be gained by going on. We have rendered a state [or] situation by coming to the end not of what is possible but of what is sensible — not of what is feasible but of what is needed. Enough is enough.[16]

> ...in actual practice we need simply proceed 'far enough'. After a certain point there is simply no need — or point — to going on.[17]

> Our explanations, interpretations, evidentiations, and substantiations can always be extended. But when we carry out these processes adequately, then after a while 'enough is enough'. The process is ended not because it has to terminate as such, but simply because there is no point in going further. A point of sufficiency has been reached. The explanation is 'sufficiently clear', the interpretation is 'adequately cogent', the evidentiation is 'sufficiently convincing'.... [T]ermination is not a matter of necessity but of sufficiency — of sensible practice rather than of inexorable principle.... What counts is doing enough 'for practical purposes'.[18]

> ...regressive viciousness in explanation can be averted ... by the consideration that the practical needs of the situation rather than considerations of general principle serve to resolve our problems here.... [I]n the end, what matters for rational substantiation is not theoretical completeness but pragmatic sufficiency.[19]

Rescher's point is a good one, and it can be buttressed by the reasoning above — certainly in the case of an endless hierarchy of probabilities. Beside practical reasons for deciding that 'enough is enough', principled considerations can be used to determine when there is a negligible difference between the value of $P(q)$ after, say, fifteen steps, or after an infinite number of them.

---

[16] Rescher 2010, 47.
[17] Ibid., 82.
[18] Rescher 2005, 104.
[19] Ibid., 105.

7.3 Rescher's Argument

Of course, it is on the basis of the context that the meaning of 'negligible' is to be understood. If one is happy to know what a particular probability is to within, say, one percent, then it is easy to work out, for given conditional probabilities, at what point the regress can be terminated, such that the error which is thereby committed is less than the desired one percent.

The third lesson, finally, must by now sound familiar: the further away A_n is from q, the smaller is the influence that the former exerts on the latter, until in the limit it dies out completely. In the end, the unconditional probabilities do not affect the value of $P(q)$ at all, only the conditional probabilities matter. Contrary to what Rescher suggests, the unconditional probability of q can be fully determined on the basis of the conditional probabilities, and of nothing else.

Again, this could be interpreted as a strengthening rather than a critique of Rescher's claims. At several places in his book Rescher explains that one of the ways in which an infinite regress can be harmless is when it is subject to "compressive convergence".[20] As he phrases it: "compressive convergence can enter in to save the day for infinite regression" (ibid.). In regresses governed by compressibility, "a law of diminishing returns" (ibid., 74) is in force, according to which the steps in the regress recede into "a minuteness of size" (ibid., 52):

> An infinite regress can thus become harmless when the regressive steps become vanishingly small in size so that the transit of regression becomes convergent. An ongoing approximation to a fixed result is then achieved, and the regress, while indeed proceeding *in infinitum*, does not reach *ad infinitum*.[21]

In the same vein, a law of diminishing returns can be said to be operating in the endless hierarchy of probabilities discussed above. Granted, it is not the case that in such a hierarchy the successive steps become smaller, let alone that they recede into "imperceptible minuteness".[22] Quite the contrary: in the limit that n goes to infinity, as has been shown, it is no impediment if $P(A_n)$ tends to the highest possible value, namely 1. Nor is it the case that, in the limit, $P(A_n)$ fades into penumbral obscurity in which its nature becomes unclear — another way in which, according to Rescher, an infinite regress can be harmless.[23] For the nature of the infinitely remote $P(A_n)$ may be perfectly clear and well-defined. Nevertheless a law of diminishing returns can still be said to be in force. Although the probability $P(A_n)$ does not shrink in

---

[20] Rescher 2010, 46.
[21] Ibid., 48.
[22] Ibid., 75.
[23] Ibid., 52.

size, nor becomes dim or otherwise unclear, the influence of $P(A_n)$ on $P(q)$, and thus the contribution that $P(A_n)$ makes to the value of $P(q)$, diminishes as the distance between A_n and q increases. This is because the hierarchical regress is isomorphic to the probabilistic regress, as we shall now prove.

7.4 The Two Regresses Are Isomorphic

In this section we will show that the regress of higher-order probabilities is strictly equivalent to the familiar probabilistic regress of propositions. Consider again (7.6). What is the second-order probability of q? It can be obtained from an instantiation of the rule of total probability at the second level:

$$\begin{aligned} P^{(2)}(q) &= P^{(2)}(q|A_1)P^{(2)}(A_1) + P^{(2)}(q|\neg A_1)P^{(2)}(\neg A_1) \\ &= \alpha_0^{(2)} v_1 + \beta_0^{(2)}(1 - v_1) \\ &= \beta_0^{(2)} + \gamma_0^{(2)} v_1, \end{aligned} \quad (7.17)$$

where v_1 was defined in (7.6), and

$$\alpha_0^{(2)} = P^{(2)}(q|A_1); \quad \beta_0^{(2)} = P^{(2)}(q|\neg A_1); \quad \gamma_0^{(2)} = \alpha_0^{(2)} - \beta_0^{(2)}. \quad (7.18)$$

According to Miller's Principle in the form (7.4), $\alpha_0^{(2)}$ is equal to v_0; but since we do not need to call on this principle for our purposes, we will let $\alpha_0^{(2)}$ stand.

The third-order probability of q is given by

$$P^{(3)}(q) = P^{(3)}(q|A_1)P^{(3)}(A_1) + P^{(3)}(q|\neg A_1)P^{(3)}(\neg A_1); \quad (7.19)$$

but the probability of A_1 at third order is no longer v_1, as it was at second order. Instead

$$\begin{aligned} P^{(3)}(A_1) &= P^{(3)}(A_1|A_2)P^{(3)}(A_2) + P^{(3)}(A_1|\neg A_2)P^{(3)}(\neg A_2) \\ &= \alpha_1^{(3)} v_2 + \beta_1^{(3)}(1 - v_2) \\ &= \beta_1^{(3)} + \gamma_1^{(3)} v_2, \end{aligned} \quad (7.20)$$

where v_2 was defined in (7.6), and

$$\alpha_1^{(3)} = P^{(3)}(A_1|A_2); \quad \beta_1^{(3)} = P^{(3)}(A_1|\neg A_2); \quad \gamma_1^{(3)} = \alpha_1^{(3)} - \beta_1^{(3)}.$$

7.4 The Two Regresses Are Isomorphic

On substituting (7.20) into Eq.(7.19) we obtain

$$P^{(3)}(q) = \alpha_0^{(3)}(\beta_1^{(3)} + \gamma_1^{(3)} v_2) + \beta_0^{(3)}(1 - \beta_1^{(3)} - \gamma_1^{(3)} v_2)$$
$$= \beta_0^{(3)} + \gamma_0^{(3)} \beta_1^{(3)} + \gamma_0^{(3)} \gamma_1^{(3)} v_2,$$

with

$$\alpha_0^{(3)} = P^{(3)}(q|A_1), \quad \beta_0^{(3)} = P^{(3)}(q|\neg A_1), \quad \gamma_0^{(3)} = \alpha_0^{(3)} - \beta_0^{(3)}, \quad (7.21)$$

which is like Eq.(7.18), except that the conditional probabilities are now at third order.

The pattern should by now be obvious. The $(m+2)$nd-order probability of q is

$$P^{(m+2)}(q) = \beta_0 + \gamma_0 \beta_1 + \gamma_0 \gamma_1 \beta_2 + \ldots + \gamma_0 \gamma_1 \ldots \gamma_{m-1} \beta_m + \gamma_0 \gamma_1 \ldots \gamma_m v_{m+1}, \quad (7.22)$$

where we have suppressed the superscript $(m+2)$ on the conditional probabilities, for reasons of legibility, but they are to be understood.

Within the usual class we obtain, in the limit that m goes to infinity,

$$P^{(\infty)}(q) = \beta_0 + \gamma_0 \beta_1 + \gamma_0 \gamma_1 \beta_2 + \gamma_0 \gamma_1 \gamma_2 \beta_3 + \ldots, \quad (7.23)$$

in which, with A_0 doing duty for the target proposition, q,

$$\alpha_n = P^{(\infty)}(A_n|A_{n+1}), \quad \beta_n = P^{(\infty)}(A_n|\neg A_{n+1}), \quad \gamma_n = \alpha_n - \beta_n, \quad (7.24)$$

for $n = 0, 1, 2, \ldots$.

It will be clear that the above argumentation on the basis of the rule of probability is formally the same as our reasoning in Chapter 3. Indeed, (7.23) has the same shape as (3.24), so an infinite series of higher-order probability statements makes sense. Like our regress of propositions that probabilistically justify one another, the regress of higher-order probabilities is subject to fading foundations and to justification that gradually emerges as we go to probability statements of higher and higher level.

Let us take stock. We have seen that higher-order probability statements are not as unintelligible as has often been thought. From Brian Skyrms and others we already learned that probabilities of the second order are not particularly problematic; but we have now seen that the same applies to probability statements of any finite order, and even that infinite-order probabilities turn out to be coherent. The two regresses, the one from the previous chapters and the hierarchical one, are formally equivalent.

However, formal equivalence is not yet equivalence in a very strict sense. We have shown that both regresses have the same form, not that there is a bijection between the two. The latter we will prove now, for the really conscientious reader.

We start straightforwardly, with the simplest form of Lewis's claim 'if something is probable, something else must be certain', i.e. the form where the series consists of only one step, namely from q to A_1. Here the two interpretations of Lewis's claim can be symbolized as follows:[24]

(1) If $P(q|A_1) = \alpha$, and $P(q|\neg A_1) < \alpha$, then A_1 is certain, i.e. $P(A_1) = 1$.
(2) It is certain that the probability of q is α, i.e. $P^{(2)}\left(P^{(1)}(q) = \alpha\right) = 1$.

It is not difficult to see that (1) entails (2). If $P(q|A_1) = \alpha$, and $P(A_1) = 1$, then

$$P(q) = P(q|A_1)P(A_1) + P(q|\neg A_1)P(\neg A_1)$$
$$= \alpha \times 1 + P(q|\neg p) \times 0$$
$$= \alpha;$$

and if $P(q) = \alpha$, which we should write more explicitly as $P^{(1)}(q) = \alpha$, then the probability that this is so is one, i.e. $P^{(2)}\left(P^{(1)}(q) = \alpha\right) = 1$.

It is a little trickier to show that (2) entails (1). The first thing we have to do is to demonstrate that $P^{(2)}\left(P^{(1)}(q) = \alpha\right) = 1$ entails $P^{(1)}(q) = \alpha$. The difficulty is that $P(A) = 1$ does not imply A in an infinite probability space. On the other hand A does entail $P(A) = 1$, so if we substitute the proposition '$P^{(1)}(q) \neq \alpha$' for A, we obtain

$$P^{(1)}(q) \neq \alpha \quad \text{entails} \quad P^{(2)}\left(P^{(1)}(q) \neq \alpha\right) = 1.$$

By contraposition it follows that

$$\neg\left[P^{(2)}\left(P^{(1)}(q) \neq \alpha\right) = 1\right] \quad \text{entails} \quad \neg[P^{(1)}(q) \neq \alpha],$$

or in other words that

$$P^{(2)}\left(P^{(1)}(q) \neq \alpha\right) \neq 1 \quad \text{entails} \quad P^{(1)}(q) = \alpha. \tag{7.25}$$

However,

$$P^{(2)}\left(P^{(1)}(q) = \alpha\right) = 1 \quad \text{implies that} \quad P^{(2)}\left(P^{(1)}(q) \neq \alpha\right) = 0,$$

---

[24] Recall that Lewis does not mention $P(q|\neg A_1)$, but we specifically include the condition of probabilistic support.

7.4 The Two Regresses Are Isomorphic

and this trivially means that $P^{(2)}\left(P^{(1)}(q) \neq \alpha\right) \neq 1$. On combining this result with (7.25), we conclude that $P^{(2)}\left(P^{(1)}(q) = \alpha\right) = 1$ entails $P^{(1)}(q) = \alpha$.[25]

The rest of the demonstration employs only first-order probabilities, so we will drop the superscript. So with $P(q) = \alpha$, and the rule of total probability,

$$P(q) = P(q|A_1)P(A_1) + P(q|\neg A_1)P(\neg A_1),$$

we see that, if A_1 is such that $P(q|A_1) = \alpha$ and $P(q|\neg A_1) < \alpha$, then

$$\alpha = \alpha \times P(A_1) + P(q|\neg A_1) \times P(\neg A_1).$$

Therefore $P(A_1) = 1$, and so we have shown that (2) entails (1).

The above shows that the two interpretations of Lewis's claim are equivalent when the series consists only of q and A_1. However, the interesting question is whether the generalization still holds when the series is longer, and especially when it is of infinite length.

The generalization of (1) and (2) above to any finite series is given by:

(1′) If $P(A_n|A_{n+1}) = \alpha_n$ and $P(A_n|\neg A_{n+1}) = \beta_n$, with $\alpha_n > \beta_n$, for $n = 0, 1, \ldots m$, then it must be that A_{m+1} is certain, i.e. $P(A_{m+1}) = 1$.

(2′) It is certain that the mth-order probability of q is v_m, i.e. $P^{(m+1)}\left(P^{(m)}\left(\ldots \left(P^{(2)}\left(P^{(1)}(q) = v_0\right) = v_1\right)\ldots\right) = v_m\right) = 1$.

We have incorporated Reichenbach's correction of Lewis's position by including β_n, i.e. the second term in the rule of total probability. The condition of probabilistic support has also been included in order to exclude multiple solutions.

We will now show that (1′) and (2′) are equivalent. The right-hand side of (7.22) matches that of (3.20) in Chapter 3, excepting only that v_{m+1} in the former replaces $P(A_{m+1})$ in the latter. But v_{m+1} is just the value of $P^{(m+2)}(A_{m+1})$, so the two equations have the same form, term for term. Going from (1′) and (2′) is immediate, whereas in the opposite direction we must first demonstrate that $P^{(m+1)}(A_m) = 1$ entails A_m. But A_m is a probability statement, so the demonstration is just the same rigmarole as the one we detailed above in going from (2) to (1). Thus the finite chains are isomorphic; and therefore, if the conditional probabilities belong to the usual class, the infinite chains have the same form too. Infinite-order probabilities are not

---

[25] A shorter, intuitive 'proof' of this result is to say that $P(B) = 1$ entails B almost everywhere, and if B is a measure, namely the proposition $P^{(1)}(q) = \alpha$, then the restriction 'almost everywhere' loses its bite.

only cogent, but they also exhibit the phenomena that we have been talking about, in particular those of fading foundations and emerging justification.

As an example of an infinite-order probability, we take as conditional probabilities

$$\alpha_n = 1 - \frac{1}{n+2} + \frac{1}{n+3} ; \qquad \beta_n = \frac{1}{n+3}.$$

These are the same as the ones we had in Eq.(3.21) of Chapter 3; but the interpretation is now different. Here they refer to infinite-order conditional probabilities. However, the equations have the same structure as those in Chapter 3; and we can read off the infinite-order probability of q by letting m go to infinity in Eq.(3.22), obtaining $P^{(\infty)}(q) = \frac{3}{4}$.

7.5 Making Coins

We have formally proved that an infinite series of higher-order probability statements is strictly equivalent to an infinite justificatory chain of the probabilistic kind. However, we might still have qualms: how can we understand the matter in an intuitive way? Being able to check all the steps in an algebraical proof is one thing, it is quite another thing to 'see through' the series, as it were, and to appreciate what is actually going on.

In this section we will try to allay these worries by offering a model that is intended to make the above abstract considerations concrete. The model is completely implementable; it comprises a procedure in which every step is specified. The model gives us a probability distribution over all the propositions as well as over their conjunctions. It satisfies the Markov condition in a very natural way, and we do not have to assume this condition as an external condition.[26]

Imagine two machines which produce trick coins. Machine V_0 produces coins each of which has bias α_0, by which we mean that each has probability α_0 of falling heads when tossed; machine W_0, on the other hand, makes coins each of which has bias β_0. We define the propositions q and A_1 as follows:

q is the proposition 'this coin will fall heads'

A_1 is the proposition 'this coin comes from machine V_0'.

---

[26] In Herzberg 2014 the Markov condition is imposed as an extra constraint. See also our discussion in Appendix A.8.

7.5 Making Coins

We shall use the symbol $P_1(q)$ for the probability of a head when A_1 is true; evidently it is the conditional probability of q, given A_1:

$$P_1(q) \stackrel{\text{def}}{=} P(q|A_1)$$
$$= P(\text{'this coin will land heads'} | \text{'this coin comes from machine } V_0\text{'}).$$

We know that $P_1(q) = \alpha_0$, for if the coin comes from machine V_0, the probability of a head is indeed α_0, for that is the bias produced by machine V_0. Note that P_1 is conceptually not the same as $P^{(1)}$. The former is a conditional probability, in this case the probability of q given A_1; the latter is a first-order unconditional probability.

An assistant is instructed to take many coins from *both* machines, and to mix them thoroughly in a large pile. The numbers of coins that she must add to the pile from machines V_0 and W_0 are determined by the properties of two new machines: V_1, which produces trick coins with bias α_1, and W_1, which produces trick coins with bias β_1. A supervisor has told the assistant that the relative number of coins that she should take from her machine V_0 should be equal to the probability, α_1, that a coin from V_1 would fall heads when tossed. So if α_1 is for example $\frac{1}{4}$, then one quarter of the total number of coins that the assistant takes from V_0 and W_0 are from V_0; the rest from W_0.[27]

The assistant takes one coin at random from her pile and she tosses it. Understanding q now to refer to this coin, we can deduce the probability of q in the new situation. Indeed, if A_2 is the proposition:

$A_2 =$ 'the relative number of V_0 coins in the assistant's pile is determined by the bias towards heads of the V_1 coins',

then we can ask what the probability is that the assistant's coin falls heads, given that A_2 is true. We use the symbol $P_2(q)$ for this probability. It is equal to the conditional probability of q, given A_2, which can be calculated from the following variation of the rule of total probability:[28]

---

[27] For the sake of this story, we limit α_1 to be a rational number, so it makes sense to say that the number of coins to be taken from V_0 is equal to α_1 times the total number taken from V_0 and W_0. Similarly, in the subsequent discussion, the biases should all be considered to be rational numbers. Since the rationals are dense in the reals, this is not an essential limitation.

[28] The proof of Eq.(7.26) goes as follows:

$$P(q \wedge A_2) = P(q \wedge A_1 \wedge A_2) + P(q \wedge \neg A_1 \wedge A_2)$$
$$= P(q|A_1 \wedge A_2)P(A_1 \wedge A_2) + P(q|\neg A_1 \wedge A_2)P(\neg A_1 \wedge A_2).$$

On dividing both sides of this equation by $P(A_2)$ we obtain (7.26).

$$P_2(q) \stackrel{\text{def}}{=} P(q|A_2) = P(q|A_1 \wedge A_2)P(A_1|A_2) + P(q|\neg A_1 \wedge A_2)P(\neg A_1|A_2).$$
(7.26)

By definition, $P(q|A_1 \wedge A_2)$ is the probability that the assistant's coin will fall heads, on condition that this coin has come from machine V_0, *and* that the number of V_0 coins in the pile is subject to the condition specified by A_2. Similarly $P(q|\neg A_1 \wedge A_2)$ is the probability that the assistant's coin will fall heads, on condition that this same coin has *not* come from machine V_0, *and* that A_2 is true.

This series of procedures gives rise to a Markov chain. For the condition that the assistant's coin has come from machine V_0 is already enough to ensure that the probability that this coin will fall heads is α_0; and that situation is not affected by the condition that A_2 is true, so $P(q|A_1 \wedge A_2) = P(q|A_1) = \alpha_0$. Likewise, the condition that the assistant's coin has not come from machine V_0 guarantees that it has come from machine W_0, and therefore ensures that the probability of a head is β_0; again, that is not affected by A_2, so $P(q|\neg A_1 \wedge A_2) = P(q|\neg A_1) = \beta_0$. In Reichenbach's locution, A_1 is said to screen off q from A_2.[29] The screening-off or Markov condition will turn out to be an essential part of our model. We shall show that the model, as well as the abstract system of which it is an interpretation, are consistent, even if the abstract system does not itself satisfy the Markov condition.

The Markov condition enables us to simplify (7.26) as follows:

$$\begin{aligned} P_2(q) = P(q|A_2) &= P(q|A_1)P(A_1|A_2) + P(q|\neg A_1)P(\neg A_1|A_2) \\ &= \alpha_0 \alpha_1 + \beta_0(1 - \alpha_1), \end{aligned}$$
(7.27)

where, as usual, we employ β_0 as shorthand for $P(q|\neg A_1)$. We conclude that, if the assistant repeats the procedure of tossing a coin from her pile many times (with replacement and randomization), the resulting relative frequency of heads would be approximately equal to $P_2(q)$, as given by (7.27). The approximation would get better and better as the number of tosses increases — more carefully: the probability that the relative number of heads will differ by less than any assigned $\varepsilon > 0$ from $\alpha_0 \alpha_1 + \beta_0(1 - \alpha_1)$ will tend to unity as the number of tosses tends to infinity.

It is important to understand that $P_2(q)$ is not simply a correction to $P_1(q)$. It is rather that they refer to two different operations. In the first operation it is certain that the assistant takes a coin from machine V_0. In the second operation something else is certain, namely that the number of V_0 coins in the pile consisting of V_0 coins and W_0 coins is determined by the bias towards heads of a coin from machine V_1. The consequence of this difference is substantial,

---

[29] Reichenbach 1956, 159-167.

7.5 Making Coins

for in the second operation it is no longer sure that the assistant takes a coin that comes from V_0. Instead of being only a correction, $P_2(q)$ is the result of a longer, and more sophisticated procedure than is $P_1(q)$.

So much for the description of the model of the first iteration of the regress, constrained by the veridicality of A_2. In the next iteration, the supervisor receives instructions from an artificial intelligence that simulates the working of yet another duo of machines, V_2 and W_2, which produce simulated coins with biases α_2 and β_2, respectively. The supervisor makes a large pile of coins from his machines V_1 and W_1; and he adjusts the relative number of coins that he takes from V_1 to be equal to the probability that a simulated coin from V_2 would fall heads when tossed. So if α_2 is for example $\frac{1}{2}$, then equal numbers of coins will be taken from each of the machines V_1 and W_1.

Let A_3 be the proposition:

$A_3 = $ 'the relative number of V_1 coins in the supervisor's pile is determined by the bias towards heads of the V_2 coins',

If A_3 is true, then the probability of A_2 is equal to α_2, that is to say $P(A_2|A_3) = \alpha_2$. Again, screening off is essential here: A_2 screens off A_1 from A_3. So we may write

$$P(A_1|A_3) = P(A_1|A_2 \wedge A_3)P(A_2|A_3) + P(A_1|\neg A_2 \wedge A_3)P(\neg A_2|A_3)$$
$$= P(A_1|A_2)P(A_2|A_3) + P(A_1|\neg A_2)P(\neg A_2|A_3)$$
$$= \alpha_1 \alpha_2 + \beta_1(1-\alpha_2). \quad (7.28)$$

This value of $P(A_1|A_3)$ is handed down to the assistant, and she reruns her procedure, but with $P(A_1|A_3)$ in place of $P(A_1|A_2)$. Since A_1 screens off q from A_3 (and from all the higher A_n), we calculate

$$P_3(q) \stackrel{\text{def}}{=} P(q|A_3) = P(q|A_1 \wedge A_3)P(A_1|A_3) + P(q|\neg A_1 \wedge A_3)P(\neg A_1|A_3)$$
$$= P(q|A_1)P(A_1|A_3) + P(q|\neg A_1)P(\neg A_1|A_3)$$
$$= \alpha_0 P(A_1|A_3) + \beta_0[1 - P(A_1|A_3)], \quad (7.29)$$

in which we are to replace $P(A_1|A_3)$ by $\alpha_1\alpha_2 + \beta_1(1-\alpha_2)$, in accordance with Eq.(7.28). This yields

$$P_3(q) = P(q|A_3) = \beta_0 + (\alpha_0 - \beta_0)\beta_1 + (\alpha_0 - \beta_0)(\alpha_1 - \beta_1)\alpha_2. \quad (7.30)$$

The relative frequency of heads that the assistant would observe will be approximately equal to $P_3(q)$, as given by (7.30) — with the usual probabilistic proviso. The above constitutes a model of the second iteration of the regress, constrained by the condition that the simulated coin of the artificial intelligence comes from the simulated machine V_2, that is by the veridicality of A_3.

This procedure must be repeated *ad infinitum*. A subprogram encodes the working of yet another duo of virtual machines, V_3 and W_3, which simulate the production of coins with biases α_3 and β_3, and so on, all under the assumption that A_n is the proposition:

$A_n = $ 'the relative number of V_{n-2} coins in the relevant pile is determined by the bias towards heads of the V_{n-1} coins'.

From this it follows that at the $(m+2)$nd step of the iteration one finds

$$P_{m+2}(q) \stackrel{\text{def}}{=} P(q|A_{m+2}) = \beta_0 + \gamma_0\beta_1 + \gamma_0\gamma_1\beta_2 \ldots + \gamma_0\gamma_1\ldots\gamma_{m-1}\beta_m$$
$$+ \gamma_0\gamma_1\ldots\gamma_m\alpha_{m+1}, \quad (7.31)$$

where we have introduced the customary abbreviation $\gamma_n = \alpha_n - \beta_n$. Under the requirement that the conditional probabilities belong to the usual class, the sequence $P_1(q), P_2(q), P_3(q)\ldots$ converges to a limit, $P_\infty(q)$, that is well-defined. Moreover, under the same condition the last term in (7.31), namely $\gamma_0\gamma_1\ldots\gamma_m\alpha_{m+1}$, tends to zero as m tends to infinity, so finally

$$P_\infty(q) = \beta_0 + \gamma_0\beta_1 + \gamma_0\gamma_1\beta_2 + \gamma_0\gamma_1\gamma_2\beta_3\ldots \quad (7.32)$$

This has the same form as (7.23).

In this way we have designed a set of procedures that is clear-cut in the sense that it could in principle be performed to any finite number of steps, where the successive results for the probability that the assistant throws a head get closer and closer to a limiting value that can be calculated. To be precise, for any $\varepsilon > 0$, and for any set of conditional probabilities that belongs to the usual class, one can calculate an integer, N, such that $|P_N(q) - P_\infty(q)| < \varepsilon$, and one could actually carry out the procedures to determine $P_N(q)$. That is, one can get as close to the limit of the infinite regress of probabilities as one likes.

The probabilities in this model are objective, but that is not the essential point. What is essential is that the structure to be described is a genuine model, which implies that two desiderata have been met. First, the model is well-defined and free from contradictions. Second, it maps into the infinite hierarchy of probabilities. The model has the same form as the probabilistic regress of Chapter 3, for which we have already given a proof of convergence. It also matches the series for the infinite-order probability of Eq.(7.23), thereby providing a model for the abstract system of Section 7.4.

7.5 Making Coins

Open Access This chapter is licensed under the terms of the Creative Commons Attribution 4.0 International License (http://creativecommons.org/licenses/by/4.0/), which permits use, sharing, adaptation, distribution and reproduction in any medium or format, as long as you give appropriate credit to the original author(s) and the source, provide a link to the Creative Commons license and indicate if changes were made.

The images or other third party material in this chapter are included in the chapter's Creative Commons license, unless indicated otherwise in a credit line to the material. If material is not included in the chapter's Creative Commons license and your intended use is not permitted by statutory regulation or exceeds the permitted use, you will need to obtain permission directly from the copyright holder.

Chapter 8
Loops and Networks

Abstract

The analysis so far concerned only one-dimensional epistemic chains. In this chapter two extensions are investigated. The first treats loops rather than chains. We show that generally, i.e. in what we have called the usual class, infinite loops yield the same value for the target as do infinite chains; it is only in the exceptional class that the values differ. The second extension involves multi-dimensional networks, where the chains fan out in many different directions. As it turns out, the uniform version of the networks yields the fractal iteration of Mandelbrot. Surprising as it may seem, justificatory systems that mushroom out greatly resemble fractals.

8.1 Tortoises and Serpents

In 1956 Wilfrid Sellars famously diagnosed the malaise of epistemology as an unpalatable either/or:

> One seems forced to choose between the picture of an elephant which rests on a tortoise (What supports the tortoise?) and the picture of a great Hegelian serpent of knowledge with its tail in its mouth (Where does it begin?). Neither will do.[1]

Up to this point our focus has been on finite and infinite chains of propositions. We looked, as it were, at an elephant which rests on a tortoise, which in turn might rest on a second tortoise, and so on, without end. *Pace* Sellars' pessimism, we have seen that such structures are not particularly problematic if one takes seriously that the relation of support is probabilistic.

---
[1] Sellars 1956, 300.

There are now two ways in which we could extend our investigation and go beyond one-dimensional chains. The first is to keep the one-dimensionality, but to look at loops rather than chains: this would take us to the second horn of Sellars's dilemma, where knowledge is pictured as Kundalini swallowing its own tail. The other way is to give up one-dimensionality altogether and to study multi-dimensional networks. This would take us to the coherentist caucus in epistemology, or rather to an infinitist version of it, in which ultimately the network stretches out indefinitely in infinitely many directions. It might seem that such a version will be especially vulnerable to the standard objection to coherentism, according to which coherentist networks of knowledge hang in the air without making contact with the world. Indeed, as Richard Fumerton noted, if we worry about "the possibility of completing one infinitely long chain of reasoning, [we] should be downright depressed about the possibility of completing an infinite number of infinitely long chains of reasoning".[2]

Remarkably enough however, the opposite is the case. Since the connections between the propositions in the network are probabilistic in character, we are dealing with conditional probabilities. As we explained in Section 4.4, the conditional probabilities together carry the empirical thrust, and this is even more so in a multi-dimensional system than in a structure of only one dimension, for the simple reason that now there are more conditional probabilities that may be linked to the world.

Extending the chains to networks thus enables us to catch it all: to develop a form of coherentism which not only is infinitist, but also acknowledges the foundationalist maxim that a body of knowledge worthy of the name must somehow make contact with the world.[3]

We start in Section 8.2 by discussing one-dimensional loops. We will see that, if justification is interpreted probabilistically, then it is in general un-

---

[2] Fumerton 1995, 57.

[3] Thus we do not have many quibbles with William Roche when he argues that foundationalism, if suitably generalized, can be reconciled with infinite regresses of probabilistic support (Roche 2016). Much depends on what is meant by foundationalism: as we indicated in Section 4.4, we do not want to become embroiled in a verbal dispute. Some commentators write as if foundationalism were the sole guardian of empirical credibility and connection to the world. Although others might find that position unduly imperialistic, we do not object to being called foundationalists in that sense. We have no issue with a form of foundationalism that takes into account fading foundations and the related concept of trading off as it is applied to doxastic justificatory chains. Our concern is less about the classification of our results than about the results themselves.

problematic to maintain that a target is justified by a loop. In Section 8.3 we turn to multi-dimensional networks, where the tentacles stretch out in many different directions. In Section 8.4 we explain that such a multi-dimensional network takes on a very interesting and intriguing shape when it goes to infinity. Surprising and somewhat strange as it may sound, if epistemic justification is interpreted probabilistically, and if we accept that it can go on without end, then justification is tantamount to constructing a fractal of the sort that Benoît Mandelbrot introduced many years ago.

In the final section we explain what happens when the multi-dimensionality springs from the *connections* in the network rather than from the nodes, i.e. when it originates from the conditional probabilities rather than from the unconditional ones. We shall see that in a generalized sense the Mandelbrot construction is preserved.[4]

8.2 One-Dimensional Loops

Finite loops embody the simplest coherentist system. What about infinite ones? It seems that an infinite loop cannot really be called a loop, since there is no end of the tail that the Hegelian serpent can swallow. A loop after all involves a repeat of the same; it may be long, indeed more than cosmologically long, but it seems that it may not be infinite, on pain of having no repetition at all. Even Henri Poincaré, when he formulated his recurrence theorem, had to assume that the universe is finite in spatial extent and of finite energy.

However, from the fact that a finite loop differs from an infinite 'loop', it does not follow that an infinite loop is in fact an infinite chain. Our investigation in this section will explain that such a conclusion would be unwarranted. In what we have called the usual class, the infinite loop indeed produces the same result as does the corresponding infinite chain; but in the exceptional class infinite loops and infinite chains yield different results, as we shall show.

We saw in Chapter 3 that the probability of the target in a finite linear chain can be written as in (3.20), where we have reinstated q in place of A_0:

$$P(q) = \beta_0 + \gamma_0\beta_1 + \gamma_0\gamma_1\beta_2 + \ldots + \gamma_0\gamma_1\ldots\gamma_{m-1}\beta_m + \gamma_0\gamma_1\ldots\gamma_m P(A_{m+1}).$$

---

[4] Section 8.2 in this chapter, about the loops, is based on Atkinson and Peijnenburg 2010a; Sections 8.3 and 8.4, which deal with networks, are based on Atkinson and Peijnenburg 2012.

The general formulation of a finite loop with $m+1$ propositions has a similar form, except that the $(m+1)$st proposition is q itself. Mathematically, there is no problem if we insert $A_{m+1} = q$ into the above equation to yield

$$P(q) = \beta_0 + \gamma_0\beta_1 + \gamma_0\gamma_1\beta_2 + \ldots + \gamma_0\gamma_1\ldots\gamma_{m-1}\beta_m + \gamma_0\gamma_1\ldots\gamma_m P(q),$$

for this yields

$$P(q) = \frac{\beta_0 + \gamma_0\beta_1 + \gamma_0\gamma_1\beta_2 + \ldots + \gamma_0\gamma_1\ldots\gamma_{m-1}\beta_m}{1 - \gamma_0\gamma_1\ldots\gamma_m}, \tag{8.1}$$

which is well-defined, on condition that $\gamma_0\gamma_1\ldots\gamma_m$ is not equal to unity.[5] With that proviso, the solution demonstrates the viability of the coherentist scenario in its simplest form, that of a finite one-dimensional loop.

The fact that a self-supporting finite loop or ring makes good mathematical sense is of course not enough. Does it also make sense elsewhere? Can a loop that closes upon itself occur in reality? A temporal example of such a loop is difficult to come by in the real world, but it can occur in the science fiction of time travel. Let q be a proposition stating that young Biff decides in 1955 to use the 2015 edition of the sports almanac, A_1 a proposition asserting that he continues his successful career as bettor until 2015, and A_2 a proposition explaining how old Biff succeeds in borrowing Doc Brown's time machine in 2015, and returns to 1955 in order to give the almanac to his younger self. $A_3 = q$ would then be a proposition stating that young Biff decides in 1955 to use the 2015 edition of the sports almanac ... and so on.

In fact, the events need not follow one another in time. Consider the following three propositions:

C: "Peter read parts of the *Critique of Pure Reason*".
P: "Peter is a philosopher".
S: "Peter knows that Kant defended the synthetic a priori".

Assuming that all philosophers read at least parts of the *Critique of Pure Reason* as undergraduates, if Peter is a philosopher, then he read parts of the *Critique*. Of course, even if he is not a philosopher, he may still have read Kant's magnum opus. If Peter knows that Kant defended the synthetic a priori, he very likely is a philosopher, whereas if he does not, he is probably not a philosopher, although of course he might be an exceptionally incompetent

---

[5] If $\gamma_0\gamma_1\ldots\gamma_m = 1$, it follows that *each* γ_n is equal to one. But then all the α_n are equal to one also, and all the β_n are equal to zero, which is the condition of bi-implication. This already indicates that a loop does not make sense when entailment relations are involved.

8.2 One-Dimensional Loops

one, not having understood anything of Kant or the *Critique*. Finally, if he read the *Critique*, he quite likely knows that Kant defended the synthetic a priori, whereas this is rather less likely if he never opened the book. Here then is a simple finite loop, consisting of a fixed number of links, namely three:

$$C \longleftarrow P \longleftarrow S \longleftarrow C, \tag{8.2}$$

where the arrow indicates that the proposition at the right-hand side probabilistically supports the one at the left.

We can make loop (8.2) nonuniform by investing the three propositions C, P and S with for example the following dissimilar values for the conditional probabilities:

C: $\alpha_0 = P(C|P) = 1$; $\quad \beta_0 = P(C|\neg P) = \frac{1}{10}$; $\quad \gamma_0 = \alpha_0 - \beta_0 = \frac{9}{10}$
P: $\alpha_1 = P(P|S) = \frac{9}{10}$; $\quad \beta_1 = P(P|\neg S) = \frac{1}{5}$; $\quad \gamma_1 = \alpha_1 - \beta_1 = \frac{7}{10}$
S: $\alpha_2 = P(S|C) = \frac{4}{5}$; $\quad \beta_2 = P(S|\neg C) = \frac{2}{5}$; $\quad \gamma_2 = \alpha_2 - \beta_2 = \frac{2}{5}$.

Then the unconditional probabilities[6] are

$$P(C) = \frac{\beta_0 + \gamma_0 \beta_1 + \gamma_0 \gamma_1 \beta_2}{1 - \gamma_0 \gamma_1 \gamma_2} = 0.711$$

$$P(P) = \frac{\beta_1 + \gamma_1 \beta_2 + \gamma_1 \gamma_2 \beta_0}{1 - \gamma_0 \gamma_1 \gamma_2} = 0.679$$

$$P(S) = \frac{\beta_2 + \gamma_2 \beta_0 + \gamma_2 \gamma_0 \beta_1}{1 - \gamma_0 \gamma_1 \gamma_2} = 0.684.$$

In the above example the number of links was fixed: there were exactly three propositions. Here is an example in which the number of links, m, can be whatever one likes, showing the cogency of *any* finite loop. Consider again the example (3.21) in Section 3.5:

$$\alpha_n = 1 - \frac{1}{n+2} + \frac{1}{n+3}; \quad \beta_n = \frac{1}{n+3}; \quad \gamma_n = 1 - \frac{1}{n+2}.$$

---

[6] As they must, these numbers satisfy

$$P(C) = \beta_0 + \gamma_0 P(P) \quad P(P) = \beta_1 + \gamma_1 P(S) \quad P(S) = \beta_2 + \gamma_2 P(C).$$

Incidentally, there is a good reason for considering a loop of at least three propositions. For in a 'loop' of two links only, there are only three independent unconditional probabilities, for example $P(q)$, $P(A_1)$ and $P(q \wedge A_1)$, whereas there are four conditional probabilities around the loop, $P(q|A_1)$, $P(q|\neg A_1)$, $P(A_1|q)$ and $P(A_1|\neg q)$, so there must be a relation between them. This difficulty does not arise for a loop of three links, for in this case there are seven independent unconditional probabilities and only six conditional probabilities around the loop. With more than three links on the loop the difference between the numbers of unconditional and conditional probabilities is even greater.

This example is nonuniform (i.e. the conditional probabilities, α_n and β_n, are not the same for different n), and it is in the usual class. It is shown in (A.18) in Appendix A.5 that Eq.(8.1) reduces to

$$P(q) = \frac{3}{4} - \frac{1}{4(m+3)}. \qquad (8.3)$$

In Table 8.1 the values of $P(q)$ for the chain are reproduced in the first line, while the corresponding values for the loop, as specified in (8.3), are given in the second line. The difference between the two cases is that, while for the chain we had to specify a value for the probability of the ground, which we put equal to a half, for the loop no such specification is required.

Table 8.1 Probability of q for chain and loop $\qquad P(p) = \frac{1}{2}$ for chain
$\alpha_n = P(A_n|A_{n+1}) = 1 - \frac{1}{n+2} + \frac{1}{n+3} \qquad \beta_n = P(A_n|\neg A_{n+1}) = \frac{1}{n+3}$

Number of A_n	1	2	5	10	25	50	75	100	∞
$P(q)$ with chain	.625	.650	.688	.712	.732	.741	.744	.745	.750
$P(q)$ with loop	.688	.700	.719	.731	.741	.745	.747	.748	.750

The probability of the target rises smoothly as the chain, or the loop, becomes longer, eventually reaching the value of three-quarters for both the infinite chain and the infinite loop. As can be seen, the values of $P(q)$ for the loop converge somewhat more quickly than do those for the chain.

The agreement between the infinite chain and the infinite loop is not limited to this special model, for it is an attribute of any example in the usual class. This can be seen quite easily, for when the product $\gamma_0 \gamma_1 \ldots \gamma_m$ tends to zero as m goes to infinity, the loop (8.1) yields the infinite, convergent series

$$P(q) = \beta_0 + \gamma_0 \beta_1 + \gamma_0 \gamma_1 \beta_2 + \gamma_0 \gamma_1 \gamma_2 \beta_3 \ldots, \qquad (8.4)$$

as for the infinite chain in the usual class.

The uniform case, in which the conditional probabilities are the same from link to link, forms an interesting special case, for then the value of $P(q)$ turns out to be always the same, no matter how many links there are in the loop. This can already be seen without doing the actual calculation. Since the propositions are uniformly connected round and round the loop *ad infinitum*, we can immediately understand why it should make no difference how many links there are: the value of $P(q)$ should be the same as that for an infinite, uniform loop. The actual calculation goes as follows: (8.1) becomes

8.3 Multi-Dimensional Networks

$$P(q) = \frac{\beta(1+\gamma+\gamma^2+\ldots\gamma^m)}{1-\gamma^{m+1}}. \tag{8.5}$$

The finite geometrical series $1+\gamma+\gamma^2+\ldots\gamma^m$ is equal to $(1-\gamma^{m+1})/(1-\gamma)$, and on substituting this we see that the factor $(1-\gamma^{m+1})$ cancels, so

$$P(q) = \frac{\beta}{1-\gamma} = \frac{\beta}{1-\alpha+\beta}.$$

Indeed this does not depend on m at all, so the number of links may be finite, or infinite, with no change in the value of $P(q)$. It will be recognized that this value is precisely the same as that for the infinite, uniform chain (see Section 3.7).

So much for the usual class. What of the exceptional class, in which the infinite product of the γ's is not zero? As we have seen, here the chain fails, in the infinite limit, to produce a definite answer for the target probability. The infinite loop on the other hand yields a unique value. To illustrate this, consider again the example (3.25):

$$\beta_n = \frac{1}{(n+2)(n+3)} \qquad \gamma_n = \frac{(n+1)(n+3)}{(n+2)^2} = 1 - \frac{1}{(n+2)^2}.$$

We find now from (8.1) that

$$P(q) = \frac{3}{4} - \frac{1}{4(m+3)}, \tag{8.6}$$

as we explain in detail in Appendix A.6, and this has the perfectly definite limit $\frac{3}{4}$. Thus the infinite chain and the infinite loop only differ in the exceptional class. There the infinite chain fails to give a definite answer, but the infinite loop does so.[7]

8.3 Multi-Dimensional Networks

Most systems of epistemic justification are of course much more complicated than the one-dimensional chains and loops that we have considered so far. Certainly modern coherentism envisages many-dimensional nets of interlocking probabilistic relations. The concept of *justification trees* or *J-trees*

---

[7] The fact that this value of $P(q)$ is the same as that of the loop (8.3), in the usual class, is just a coincidence.

has been introduced as a graphic representation of the relation in such networks.[8] Figure 8.1 is an example of a very simple justification tree. This tree has two branches, with A_1 and A'_1 as nodes on the one level, and A_2 and A'_2 as nodes on a lower level. It should be read as: proposition q is justified by A_1 and A'_1, A_1 is justified by A_2, and A'_1 is justified by A'_2. In this section we shall describe what happens when we replace the finite or infinite one-dimensional probabilistic chain by a finite or infinite probabilistic network in two dimensions, along the lines of a justification tree.

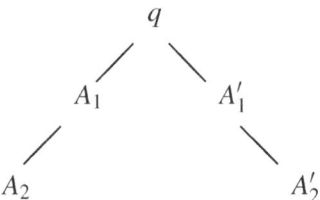

Fig. 8.1 Basic justification tree

We now make the tree more complicated by allowing that A_1 and A'_1 are each supported by two, rather than by one proposition, as depicted in Fig. 8.2.

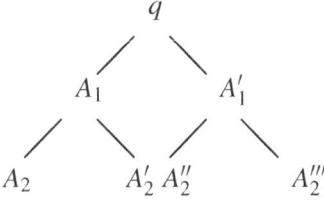

Fig. 8.2 Complex justification tree

Here A_1 is supported by A_2 and A'_2; and A'_1 is supported by A''_2 and A'''_2. In their turn, A_2, A'_2, A''_2, and A'''_2 may each be supported by two propositions.

A complicated tree as in 8.2 could serve as a model for the propagation of genetic traits under sexual reproduction, in which the traits of a child

---
[8] See for example Sosa 1979; Clark 1988, 374-375; Alston 1989, 19-38; Cortens 2002, 25-26; Aikin 2011, 74.

8.3 Multi-Dimensional Networks

are related probabilistically to those of both the mother and the father. Let $P(q)$ again be the unconditional probability that Barbara has trait T. This time Barbara is not a bacterium as in Section 3.7, where the reproduction was asexual. Rather she is now an organism with two parents, a father and a mother. For the purpose of fixing ideas it will prove convenient to talk about sexual reproduction and about fathers and mothers, but we should bear in mind that the formalism is of course much more general. Also, although we shall tell the story in terms of events, it should be kept in mind that everything we say applies to justificatory relations between propositions as well.

Since Barbara stems from two parents, the probability that she has T is determined by the characteristics of her mother and of her father. Rather than two reference classes (the mother having or not having T), we now have four: both the mother and the father have T, neither of them has it, the father has T but the mother does not, and the mother has T but the father does not. The analogue of the rule of total probability is

$$P(q) = \alpha_0 P(A_1 \wedge A'_1) + \beta_0 P(\neg A_1 \wedge \neg A'_1)$$
$$+ \gamma_0 P(A_1 \wedge \neg A'_1) + \delta_0 P(\neg A_1 \wedge A'_1), \quad (8.7)$$

where A_1 represents Barbara's mother having T and A'_1 her father having T. Here α_0 means "the probability that Barbara has T, given that her mother and father both have T". The other conditional probabilities are analogously defined: β_0 corresponds to neither parent having T, and γ_0 and δ_0 to the two situations in which one parent does, and the other does not have T.

In the nth generation the corresponding expression is

$$P(A_n) = \alpha_n P(A_{n+1} \wedge A'_{n+1}) + \beta_n P(\neg A_{n+1} \wedge \neg A'_{n+1})$$
$$+ \gamma_n P(A_{n+1} \wedge \neg A'_{n+1}) + \delta_n P(\neg A_{n+1} \wedge A'_{n+1}), \quad (8.8)$$

where A_n stands for one individual in the nth generation, A_{n+1} and A'_{n+1} for that individual's mother and father. The conditional probabilities are

$$\alpha_n = P(A_n | A_{n+1} \wedge A'_{n+1})$$
$$\beta_n = P(A_n | \neg A_{n+1} \wedge \neg A'_{n+1})$$
$$\gamma_n = P(A_n | A_{n+1} \wedge \neg A'_{n+1})$$
$$\delta_n = P(A_n | \neg A_{n+1} \wedge A'_{n+1}).$$

In order to iterate the two-dimensional (8.8), much as we did in the one-dimensional case, we now need more complicated relations for the unconditional probabilities. It is no longer sufficient to consider $P(A_1)$ and replace it by $\beta_0 + (\alpha_0 - \beta_0)P(A_2)$, and so on, for now we are dealing with the probability of a conjunction of two parents, A_1 and A'_1. Each of these parents has two parents, so we encounter in fact the probabilities of conjunctions of four individuals. This can be continued further and further, involving more and more progenitors, confronting us with a tree of increasing complexity.

Fortunately, however, we can often make simplifying assumptions. Here we will work under three simplifications:

1. **Independence.** The probabilities for the occurrence of the trait T in females and in males is independent of one another in any of the n generations:

$$P(A_{n+1} \wedge A'_{n+1}) = P(A_{n+1})P(A'_{n+1}).$$

This assumption seems reasonable in the genetic context; and it will also apply in many more general epistemological settings.

2. **Gender symmetry.** The probability of the occurrence of the trait T is the same for females and for males in any of the n generations:

$$P(A_n) = P(A'_n).$$

Thus we only consider inheritable traits which are gender-independent, such as having blue eyes or being red-haired, and not, for example, having breast cancer or being taller than two metres. Similarly, in an epistemological context this assumption will sometimes, but not always be satisfied. With this assumption the prime can be dropped on A'_n, and in combination with the first assumption we obtain

$$P(A_{n+1} \wedge A'_{n+1}) = P(A_{n+1})P(A_{n+1}) = P^2(A_{n+1}).$$

3. **Uniformity.** The conditional probabilities are the same in any of the n generations. That is, α_n, β_n, γ_n and δ_n are independent of n, so we may drop the suffix.

Together these assumptions enable us to simplify (8.8) to the quadratic function

$$P(A_n) = \alpha P^2(A_{n+1}) + \beta P^2(\neg A_{n+1}) + (\gamma + \delta)P(A_{n+1})P(\neg A_{n+1}). \quad (8.9)$$

As we will show in the next section, (8.9) leads to a surprising result, for it generates a structure similar to the Mandelbrot fractal.

8.4 The Mandelbrot Fractal

In 1977 Mandelbrot introduced his celebrated iteration:

$$q_{n+1} = c + q_n^2, \qquad (8.10)$$

where c and q are complex numbers.[9] Starting with $q_0 = 0$, the iteration gives us successively

$$\begin{aligned} q_1 &= c \\ q_2 &= c + c^2 \\ q_3 &= c + (c + c^2)^2 \\ q_4 &= c + \left(c + (c + c^2)^2\right)^2, \end{aligned} \qquad (8.11)$$

and so on. For many values of c, the iteration will diverge, allowing q_n to grow beyond any bound as n becomes larger and larger. For example, if $c = 1$ we obtain $q_1 = 1$, $q_2 = 2$, $q_3 = 5$ and $q_4 = 26$, and so on.

But if for instance $c = 0.1$, then q_n does not diverge, and in this case actually converges to the number $0.11271\ldots$. Taken together, all the values of c for which the iteration (8.10) does not diverge form the Mandelbrot set, which is reproduced in Figure 8.3.

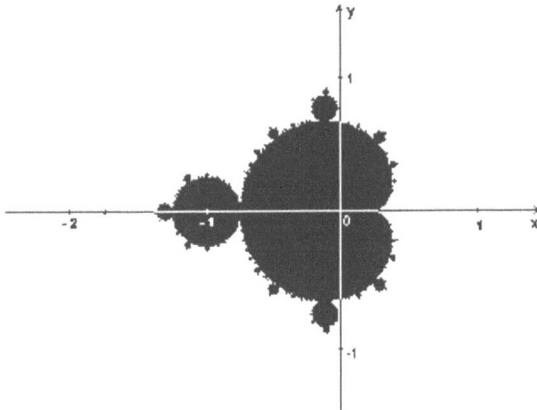

Fig. 8.3 The Mandelbrot fractal is generated by the complex quadratic iteration $q_n = c + q_{n+1}^2$, where $c = x + iy$.

---

[9] Mandelbrot 1977. The variables q_n in this section should not be confused with q in (8.7), the target proposition of the two-dimensional net.

The black area contains the points that belong to the Mandelbrot set. Each point corresponds to a complex number, c, being the ordered pair of the Cartesian coordinates, (x,y). The edge of the Mandelbrot set forms the boundary between those values of c that are members of the set and those that are not. It is this boundary, the 'Mandelbrot fractal', that has the well-known property of being infinitely structured in a remarkable way: no matter how far you zoom in on it, you will always find a new structure that is similar to, although not completely identical with the Mandelbrot set itself.

Our aim in this section is to demonstrate that, on condition that $\alpha + \beta \neq \gamma + \delta$, the quadratic relation (8.9) is equivalent to the Mandelbrot iteration (8.10). As it turns out, c will be a function of the conditional probabilities α, β, γ and δ alone, and will thus be a known quantity. The q's, on the other hand, will be directly related to the unconditional probabilities; these are unknown and their values are to be determined through the iteration.

It will prove convenient first to define ε as the average of the conditional probabilities γ and δ, that is

$$\varepsilon \stackrel{\text{def}}{=} \tfrac{1}{2}(\gamma + \delta),$$

which is the mean conditional probability that the target — in our case Barbara — has the trait T, given that only one of her parents has T. Eq.(8.9) now becomes

$$P(A_n) = \beta + 2(\varepsilon - \beta)P(A_{n+1}) + (\alpha + \beta - 2\varepsilon)P^2(A_{n+1}). \quad (8.12)$$

On the one hand, this iteration may not look very much like the Mandelbrot form (8.10). Firstly, in the latter we go as it were upwards, starting from q_n and then counting to q_{n+1}, whereas in (8.12) we start with $P(A_{n+1})$ and iterate downwards to $P(A_n)$. Secondly, (8.12) is about conditional and unconditional probabilities, and thus about real numbers between zero and one, whereas (8.10) is an uninterpreted formula involving complex numbers. On the other hand, however, we see that there is an important similarity between (8.10) and (8.12). Both are quadratic expressions: the former contains q_n^2 and the latter $P^2(A_{n+1})$. In order to transform (8.12) into (8.10) we introduce a linear mapping that serves to remove from (8.12) the term $2(\varepsilon - \beta)P(A_{n+1})$, and also the coefficient $(\alpha + \beta - 2\varepsilon)$. The appropriate linear mapping that does the trick, $P(A_n) \to q_n$, is defined by

$$q_n = (\alpha + \beta - 2\varepsilon)P(A_n) - \beta + \varepsilon. \quad (8.13)$$

On substituting (8.12) for $P(A_n)$ in (8.13) we obtain a formula that can be rewritten as

8.4 The Mandelbrot Fractal

$$q_n = \varepsilon(1-\varepsilon) - \beta(1-\alpha) + q_{n+1}^2. \tag{8.14}$$

The details of this calculation can be found in Appendix D.2.

Now define

$$c = \varepsilon(1-\varepsilon) - \beta(1-\alpha). \tag{8.15}$$

Note that c involves only the conditional probabilities, α, β and ε, and so is an invariant quantity during the execution of the iteration. On the other hand, q_n also contains the unconditional probability, $P(A_n)$, which we seek to evaluate through the iteration. With the definition (8.15), Eq.(8.14) becomes

$$q_n = c + q_{n+1}^2. \tag{8.16}$$

Evidently (8.16) is very similar to the standard Mandelbrot iteration (8.10). There is only the one difference which we have already mentioned: instead of an iteration upwards from $n = 0$, the iteration in (8.16) proceeds from a large n value, corresponding to the primeval parents, down to the target child proposition at $n = 0$. This difference is however only cosmetic and has no significance for the iteration as such.

We are now in a position to take advantage of some of the lore that has accumulated about the Mandelbrot iteration. *Some* but not all, for there is still the second difference that we mentioned: epistemic justification as we discuss it here deals with probabilities, and those are real numbers, rather than complex ones. Hence we must concentrate on the real subset of the complex numbers c in (8.15), namely those for which $c = (x, 0)$, corresponding to the x-axis in Figure 8.3. It should be noted that, when c is real, all the q_n are automatically real — compare the explicit expressions for the first few n-values, just after (8.11). It is known that the real interval $-2 \leq c \leq \frac{1}{4}$ lies within the Mandelbrot set, but not all of these values correspond to an iteration that converges to a unique limiting value.

However, let us now impose the condition of probabilistic support, with exclusion of zero and one. Although $0 < \beta < \alpha < 1$ has the same form as the condition of probabilistic support for the one-dimensional chain, it should be realized that α and β do not have quite the same meanings in the two contexts. In the one-dimensional chain, $\alpha > \beta$ means that the probability of the child's having trait T is greater if the mother has it than if the mother does not have it. In the two-dimensional net, however, $\alpha > \beta$ means that the probability of the child's having trait T is greater if *both* of her parents have it than if *neither* of them do.

The essential point is that with $0 < \beta < \alpha < 1$ we can show from (8.15) that $-\frac{1}{4} < c < \frac{1}{4}$ (see again Appendix D.2). In this domain the Mandelbrot iteration is known to converge to a unique limit. Were it not for probabilistic support, convergence would not be guaranteed, indeed a so-called two-cycle, in which q_n flips incessantly between two values, would have been a possibility. Hence the condition of probabilistic support is necessary for convergence in this case.

A fixed point of the mapping (8.16) is a number, q_*, that satisfies

$$q_* = c + q_*^2. \tag{8.17}$$

In Appendix D it is proved that the solution

$$q_* = \frac{c}{\frac{1}{2} + \sqrt{\frac{1}{4} - c}}, \tag{8.18}$$

is the so-called attracting fixed point of (8.16), meaning that the iteration (8.16) converges to q_*. Independently of the value one takes as the starting point for the iteration (i.e. q_N for some large N), attraction to the same q_* takes place (on condition that the starting point is not too far from q_* — technically, the condition is that it is within the basin of attraction of the fixed point). Under these conditions the starting point or ground has no effect on the final value of the target, q_0. The phenomenon is precisely that of fading foundations, now in the context of a two-dimensional net.

This fixed point (8.18) corresponds to the following fixed point of (8.12):

$$p_* = \frac{\beta}{\beta + \frac{1}{2} - \varepsilon + \sqrt{\beta(1 - \alpha) + (\varepsilon - \frac{1}{2})^2}}. \tag{8.19}$$

Note that, if $\varepsilon = \frac{1}{2}(\alpha + \beta)$, which is equivalent to $\alpha + \beta = \gamma + \delta$, p_* reduces to $\beta/(1 - \alpha + \beta)$, and this agrees with the sum of the one-dimensional iteration (3.17).

If β tends to zero the solution (8.19) is interesting, for it vanishes only if $\varepsilon \leq \frac{1}{2}$. If $\varepsilon > \frac{1}{2}$ it tends to the nontrivial value $(2\varepsilon - 1)/(2\varepsilon - \alpha)$ — see Appendix D.2. This behaviour is different from that of the one-dimensional case, in which the solution always vanishes when β tends to zero.

The two-dimensional network is generated by the same recursion that produces the Mandelbrot set in the complex plane. True, we have only to do with the real line between $-\frac{1}{4}$ and $\frac{1}{4}$, and not with the complex plane (where the remarkable fractal structure is apparent). But the point is that the algorithm which produces our sequence of probabilities, and that which generates the Mandelbot fractal, are the same.

We have used three simplifying assumptions in proving the above properties, viz. those of independence, probabilistic symmetry between A_{n+1} and

8.4 The Mandelbrot Fractal

A'_{n+1}, and uniformity. There are however strong indications that essentially similar results also hold when these assumptions are dropped. Imagine a situation in which the probabilities are different for A_{n+1} and A'_{n+1}. Then there will be two coupled quadratic iterations, one for $P(A_n)$ and one for $P(A'_n)$. Each of these is related to $P(A_{n+1})$ as well as $P(A'_{n+1})$. This is however merely a technical complication, for it is still possible to find a domain in which the iterations converge. The relation is in fact a generalized Mandelbrot iteration, and analogous results obtain.

The same applies if we drop the assumption of independence. Clearly, if A_{n+1} and A'_{n+1} are stochastically dependent, we may have to include more distant links in the network, which of course complicates matters considerably. However, in general terms it means nothing more than that the final fixed-point equations will be of higher order. Again a generalized Mandelbrot-style iteration will hold sway, and again domains of convergence will exist.

Furthermore, in many situations the conditional probabilities may not be uniform: they may change from generation to generation. In those cases the iteration will become considerably more involved. We have seen that for the one-dimensional chain it proved possible to write down explicitly the result of concatenating an arbitrary number of steps. It is true that for a two-dimensional net this would be very cumbersome. However, with the use of a fixed-point theorem it is possible to give conditions under which convergence once more occurs.

What will happen when the network has more dimensions than two? In that case the fixed-point equations will be of even higher order, necessitating computer programs for their calculation. The picture itself however remains essentially the same. The probabilities are determined by polynomial recurrent expressions, and there will be a domain in which they are uniquely determined.

We conclude that probabilistic epistemic justification has a structure that gives rise to a generalized Mandelbrot recursion. This still holds when we abandon our three simplifying assumptions, or when we work in more than two dimensions. In short, not only do the algorithms describing ferns, snowflakes and many other patterns in nature generate a fractal, but the same is true for the description of our patterns of reasoning.

8.5 Mushrooming Out

Consider once more our justificatory chain in one dimension

$$q \longleftarrow A_1 \longleftarrow A_2 \longleftarrow A_3 \longleftarrow A_4 \ldots$$

where the arrow is again interpreted as probabilistic support. Above we have constructed multi-dimensional networks by letting new chains spring from the nodes, that is the unconditional probabilities. However chains can also arise from the connections, that is from the arrows. This possibility seems to be have been anticipated by Richard Fumerton.

Fumerton has observed that many examples of sceptical reasoning rely on a principle which he calls the Principle of Inferential Justification. The principle consists of two clauses:

> To be justified in believing one proposition q on the basis of another proposition A_1, one must be (1) justified in believing A_1 and (2) justified in believing that A_1 makes probable q.[10]

He then argues that, ironically, the same principle is used to *reject* scepticism and to support classic foundationalism:

> The foundationalist holds that every justified belief owes its justification ultimately to some belief that is noninferentially justified. ...The principle of inferential justification plays an integral role in the famous regress argument for foundationalism. If all justification were inferential, the argument goes, we would have no justification for believing anything whatsoever. If all justification were inferential, then to be justified in believing some proposition q I would need to infer it from some other proposition A_1. According to the first clause of the principle of inferential justification, I would be justified in believing q on the basis of A_1 only if I were justified in believing A_1. But if all justification were inferential I would be justified in believing A_1 only if I believed it on the basis of something else A_2, which I justifiably believe on the basis of something else A_3, which I justifiably believe on the basis of something else A_4, ..., and so on ad infinitum. Finite minds cannot complete an infinitely long chain of reasoning, so if all justification were inferential we would have no justification for believing anything.[11]

---

[10] Fumerton 1995, 36; 2001, 6. We have substituted q and A_1 for Fumerton's P and E. Fumerton applies the principle in particular to scepticism of what he calls the "strong" and "local" kind (Fumerton 1995, 29-31). Strong scepticism denies that we can have justified or rational belief; it is opposed to weak scepticism, which denies that we can have knowledge. Local scepticism is scepticism with respect to a given class of propositions, whereas global scepticism denies that we can know or rationally believe *all* truth.

[11] Fumerton 1995, 56-57.

8.5 Mushrooming Out

We recognize here the finite mind objection to infinite justificatory chains, which we discussed in Chapter 5. This objection, that serves as an argument in support of foundationalism, alludes to the first clause of the Principle of Inferential Justification, and it consitutes the first part of Fumerton's epistemic regress argument for foundationalism.[12] There is however a second part to Fumerton's epistemic regress argument. This part depends on the second clause of the Principle of Inferential Justification, and it has to do with multi-dimensionality arising from chains that spring from connections rather than from nodes. Here again an infinite number of infinite regresses mushroom out in infinitely many directions:

> To be justified in believing q on the basis of A_1, we must be justified in believing A_1. But we must also be justified in believing that A_1 makes probable q. And if all justification is inferential, then we must justifiably infer that A_1 makes probable q from some proposition B_1, which we justifiably infer from some proposition B_2, and so on. We must also justifiably believe that B_1 makes probable that A_1 makes probable q, so we would have to infer that from some proposition C_1, which we justifiably infer from some proposition C_2, and so on. And we would have to infer that C_1 makes probable that B_1 makes probable that A_1 makes probable q ... The infinite regresses are mushrooming out in an infinite number of different directions.[13]

The consequences of this particular mushrooming out seem to be bleak indeed, as Fumerton notes:

> If finite minds should worry about the possibility of completing one infinitely long chain of reasoning, they should be downright depressed about the possibility of completing an infinite number of infinitely long chains of reasoning.[14]

Fortunately, however, things are not as grim as Fumerton suggests. The situation is on the contrary very interesting. For Fumertonian mushrooming out generates a Mandelbrot-like iteration of the sort that we described in the previous section.

Let us explain. In the previous chapters we have thought of the conditional probabilities as somehow being given: they were measured or estimated, for

---

[12] For Fumerton's distinction between the *epistemic* and the *conceptual* regress argument for foundationalism, see Section 6.1. There we argued that the conceptual regress argument amounts to the no starting point objection to infinite epistemic chains.
[13] Fumerton 1995, 57. B_1, C_1 etc. come in the place of Fumerton's F_1, G_1.
[14] Ibid.

instance in a laboratory, as in our example about the bacteria. With given conditional probabilities, there is of course no Fumertonian mushrooming out: we can iterate the unconditional probabilities in the usual way on the basis of the conditional probabilities as our pragmatic starting point. However, Fumerton is right to intimate that sometimes the conditional probabilities are unknown or at least uncertain; then their values have to be justified by some further proposition, which has to be justified by yet another proposition, and so on, and we are faced with mushrooming in Fumerton's sense. How to deal with this situation?

Again let q be probabilistically supported by A_1:

$$P(q|A_1) > P(q|\neg A_1).$$

Now suppose that these two conditional probabilities are not given. The only thing we know is that "q is probabilistically supported by A_1" is in turn made probable by another proposition, for example by B_1. The way to express this is by writing down the relevant rules of total probability, this time for conditional rather than unconditional probabilities:

$$P(q|A_1) = P(q|A_1 \wedge B_1)P(B_1|A_1) + P(q|A_1 \wedge \neg B_1)P(\neg B_1|A_1) \quad (8.20)$$
$$P(q|\neg A_1) = P(q|\neg A_1 \wedge B_1)P(B_1|\neg A_1) + P(q|\neg A_1 \wedge \neg B_1)P(\neg B_1|\neg A_1).$$

These rules are clearly more complicated than the simple rule for an unconditional probability, although we already encountered this complicated form in (7.26) of Chapter 7, when we discussed our model for higher-order probabilities.[15]

The unconditional probability $P(q)$ can be written as

$$P(q) = P(q|A_1)P(A_1) + P(q|\neg A_1)P(\neg A_1),$$

and on using (8.20) to evaluate the two conditional probabilities, we find that

$$P(q) = \left[P(q|A_1 \wedge B_1)P(B_1|A_1) + P(q|A_1 \wedge \neg B_1)P(\neg B_1|A_1)\right]P(A_1)$$
$$+ \left[P(q|\neg A_1 \wedge B_1)P(B_1|\neg A_1) + P(q|\neg A_1 \wedge \neg B_1)P(\neg B_1|\neg A_1)\right]P(\neg A_1)$$
$$= \alpha_0 P(A_1 \wedge B_1) + \gamma_0 P(A_1 \wedge \neg B_1) + \delta_0 P(\neg A_1 \wedge B_1) + \beta_0 P(\neg A_1 \wedge \neg B_1).$$

The last line has precisely the structure of (8.7), reading B_1 here for A'_1 there. This shows that a single mushrooming out à la Fumerton is isomorphic to the two-dimensional equations of the previous section.

---

[15] An intuitive way of seeing that (8.20) is correct is to realize that, in the reduced probability space in which A_1 is the whole space, all the occurrences of A_1 can be omitted. Then (8.20) reduces to the rule of total probability for an unconditional probability.

8.6 Causal Graphs

We have seen that, where chains spring from the nodes, the two-dimensional equations could be extended to equations in many, and even infinitely many dimensions, yielding a Mandelbrot structure. The same reasoning can be applied here, where chains spring from the connections. If many, or even a denumerable infinity of conditional probabilities are in turn probabilistically supported, then one has to do with the many-dimensional generalization.

Of course we will never deal with all these dimensions in reality. Our result is first and foremost a formal one. Having said this we should not underestimate the relevance of formal results for real life justification. Although it is true that in justifying our beliefs we can handle only short, finite chains, it is thanks to formal reasoning that we can recognize in these chains the manifestation of fading foundations: solely through formal proofs do we know that what we see in real life justification is not a fluctuation or a coincidence.[16]

8.6 Causal Graphs

In the first chapter we briefly referred to the similarities between epistemic and causal chains. Especially at a formal level, as we stressed in Chapter 2, a chain of reasons and a chain of causes are very much alike. Thus the linear chain

$$A_0 \longleftarrow A_1 \longleftarrow A_2 \longleftarrow A_3 \longleftarrow A_4 \longleftarrow \ldots \qquad (8.21)$$

can be interpreted as a one-dimensional causal series, where A_0 is the fact or event (rather than the proposition) that bacterium Barbara from Chapter 3 has trait T, and A_1 is the fact or event that her mother had T, and so on, backwards in time. The arrows in (8.21) stand for probabilistically causal influences: if a mother has T, it is more likely, but not certain, that her daughter will have T. This is in line with ordinary usage, for example when one says that smoking causes lung cancer, even though one knows that not all smokers contract the affliction, and that some non-smokers succumb to it. To avoid cumbersome language, we shall sometimes say that A_0 stands for Barbara

---

[16] As the size and complexity of the multi-dimensional networks increase, it will become more and more difficult to have them correspond to empirically based conditional probabilities. A rather wild speculation is that in the end such a world-network might have only one solution. See Atkinson and Peijnenburg 2010c, where we mull over the implications of such a speculation, taking as our starting point Susan Haack's crossword metaphor for 'foundherentism' (Haack 1993, Chapter 4).

(rather than for the fact that Barbara has T), that A_1 stands for her mother (rather than for the fact that her mother has T), and so on.

In the language of Directed Acyclic Graphs (DAGs) one would say that (8.21) is a DAG just in case the Markov condition holds.[17] This means in particular that A_1 screens off A_0 from A_2 in the sense of Reichenbach, that A_2 screens off A_1 from A_3, and so on.[18] However, the Markov condition is much stronger than a screening-off constraint that involves only three successive events. The idea is that the 'parent event' of a 'child event' screens off the child from any and all 'ancestor events', or combinations thereof. For the chain of (8.21), the condition is formally as follows:

$$P(A_n|A_{n+1} \wedge Z) = P(A_n|A_{n+1})$$
$$P(A_n|\neg A_{n+1} \wedge Z) = P(A_n|\neg A_{n+1}),$$

for all $n \geq 0$. Here Z stands for any event, A_m, in the chain, apart from the descendents of A_n, i.e. for any $m \geq n+2$, or for any conjunction of such events, or their negations. This can be written succinctly as

$$P(A_n|\pm A_{n+1} \wedge Z) = P(A_n|\pm A_{n+1}),$$

where it is understood that $+A_{n+1}$ simply means A_{n+1}, and $-A_{n+1}$ means $\neg A_{n+1}$. The idea, informally, is that the Markov condition ensures that the causal influences which probabilistically circumscribe Barbara's genetic condition are determined by her mother alone, and that one can forget about all her ancestors except for her mother.

It should be stressed that our analysis of the probabilistic regress in no way requires the imposition of the Markov condition: fading foundations and the emergence of justification in the case of a justificatory regress work just as well with, as without the Markov condition. The causal influence of the primal ancestor fades away as the distance between Barbara and the ancestor increases, and Barbara's probabilistic tendency to have T emerges from the causal regress, whether or not the Markov condition holds.

It is certainly possible, in a particular causal chain, that fact A_2 could have a causal influence on A_0 directly, apart from its indirect influence through A_1. Hesslow has given an example.[19] Birth control pills, A_2, directly increase the probability of thrombosis, A_0, but indirectly reduce it in sexually active women by reducing the probability of pregnancy, A_1, which itself constitutes

---

[17] Spirtes, Glymour and Scheines 1993; Pearl 2000; Hitchcock 2012.
[18] Reichenbach 1956.
[19] Hesslow 1976.

8.6 Causal Graphs

a thrombosis risk. Then the Markov condition, as we have stated it for (8.21), would break down, and one would have to add a direct causal link between A_0 and A_2, as shown in Figure 8.4. In this case a modified Markov condition could still be in force: now both A_1 and A_2 count as parent events of A_0, and they together might screen off A_0 from the rest of the chain (depending on the details of the case, of course).

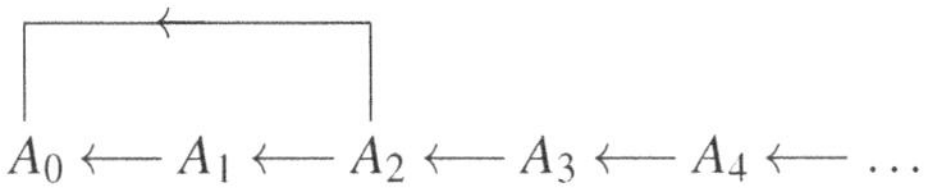

$$A_0 \longleftarrow A_1 \longleftarrow A_2 \longleftarrow A_3 \longleftarrow A_4 \longleftarrow \ldots$$

Fig. 8.4 Modified causal chain

An advantage of the above considerations concerning the Markov condition is that they facilitate a demonstration of the consistency of our probabilistic regress.[20] This works just as well for the regress of justification as it does for the regress of causes. The idea is that, with the Markov condition in place, one can work out the probabilities of the conjunction of any of the A_n in terms of the usual conditional probabilities and the unconditional probabilities of the A_n, which, as we know, can be calculated from the conditional probabilities alone (on condition of course that the latter are in the usual class). For example, as shown in Appendix A.8,

$$P(A_1 \wedge \neg A_3 \wedge A_4) = (\beta_1 + \gamma_1 \beta_2)(1 - \alpha_3) P(A_4).$$

So there is a probability distribution over all the conjunctions of events (or propositions), and thus the probabilistic regress is consistent in this sense. If the Markov constraint is not imposed, on the other hand, so that the chain may not be a genuine DAG, then there are in general many ways to distribute probabilities over the various conjunctions; but we are sure that there is at least one way, thanks to Markov, that is consistent.

Let us now progress from one to two dimensions. Consider the tree 8.2 of Section 8.3, but now reinterpreted as a causal net:
Note that, while the direction of epistemic support in Figure 8.2 is from the bottom of the figure to the top, the direction of causal influence in Figure 8.5 is from top to bottom. Thus event q probabilistically causes events A_1 and A'_1, and A_1 in turn causes A_2 and A'_2, while A'_1 causes A''_2 and A'''_2. For example, q could stand for Barbara's grandmother — more accurately, for the event that Barbara's grandmother had T. Through binary fission this grandmother

---
[20] Herzberg 2013.

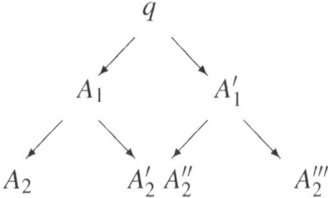

Fig. 8.5 Two-dimensional causal net with common causes

would split into two daughter cells, which would probably, but not certainly, have T. Then A_1 could stand for Barbara's mother, and finally A_2 for Barbara herself, A_2' for her sister bacterium. The eventualities A_1', A_2'' and A_2''' would have analogous meanings in respect of Barbara's aunt and her cousins.

One would expect the following Markov condition to hold, namely that A_1 screens off A_2 and A_2' from all the other events in the net. Thus

$$P(A_2|\pm A_1 \wedge Z) = P(A_2|\pm A_1)$$
$$P(A_2'|\pm A_1 \wedge Z) = P(A_2'|\pm A_1),$$

where Z can be any of q, A_1', A_2'' or A_2''', or their negations, or any conjunctions of the same. Similarly, A_1' screens off A_2'' and A_2''' from q, A_1, A_2 and A_2'. One would also expect A_2 and A_2' to be positively correlated, so

$$P(A_2 \wedge A_2') > P(A_2)P(A_2'),$$

although they are conditionally independent in the sense that

$$P(A_2 \wedge A_2'|\pm A_1) = P(A_2|\pm A_1)P(A_2'|\pm A_1).$$

This equation is in fact a consequence of the Markov condition. Following Reichenbach, we say that A_1 is the common cause of A_2 and A_2', and that event A_1 has brought it about that A_2 is more likely to occur if A_2' occurs, and *vice versa*.

A different kind of causal net is shown in Figure 8.6. Here the causal arrows go from bottom to top, which is the same as the direction of epistemic support in Figure 8.2. In Figure 8.6, A_2 could stand for a mother (i.e. for the event that a mother carries a particular trait, for example having blue eyes), A_2' could stand for her husband, and A_1 could stand for their daughter. Assuming that mother and father were not related, A_2 and A_2' are unconditionally independent,

$$P(A_2 \wedge A_2') = P(A_2)P(A_2'),$$

8.6 Causal Graphs

but they become correlated on conditionalization by A_1,

$$P(A_2 \wedge A'_2 | \pm A_1) \neq P(A_2 | \pm A_1) P(A'_2 | \pm A_1).$$

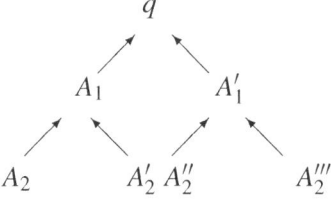

Fig. 8.6 Two-dimensional causal net with unshielded colliders

The subgraph involving A_2, A'_2 and A_1 is a so-called unshielded collider. The behaviour of this collider, insofar as conditional and unconditional dependencies are concerned, is just the opposite of the behaviour of the common cause. Clearly Figure 8.6 is more like the two-dimensional justification tree of 8.2 than is the common cause graph of Figure 8.5. In the justification tree, proposition A_1 is probabilistically supported by A_2 and A'_2: moieties of justification accrue to A_1 from A_2 and A'_2, and from the conditional probabilities. In the causal collider, A_1 is probabilistically caused by A_2 and A'_2. Similarly, parents A''_2 and A'''_2 cause A'_1, the event that their son carries the trait in question. And finally A_1 and A'_1 can cause the event that a child in the third generation has blue eyes.

Strictly speaking, Figure 8.6 is inaccurate, or at least ambiguous. The point is that A_1 would not be caused at all by A_2 in the absence of A'_2. We should replace Figure 8.6 by Figure 8.7, in which the joint nature of the causal influences is explicitly represented.

Mathematically, such a picture is called a directed hypergraph; and its properties have been studied by Selim Berker in the context of justificatory trees rather than causal trees.[21] Berker makes the point that such hypergraphs offer coherentists and infinitists a way of attaching a justification tree of beliefs or propositions to empirical facts. This is done without thereby making them foundational trees in which the facts constitute grounds in the sense of the foundationalist, that is as regress stoppers. For example, suppose now that A_2 in Figure 8.7 is an agent's experience that the sun is shining, and that

---

[21] Berker 2015.

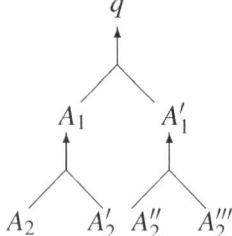

Fig. 8.7 Two-dimensional hypergraph

A'_2 is her belief that her eyes and visual cortex are functioning normally. Then A_1 could be the belief that the sun is indeed shining. The crux of the matter is that the fact A_2 does not by itself justify A_1, but does so only together with A'_2.

Berker claims that a coherentist (or infinitist) account of justification cannot consistently be based on probabilistic considerations. His reasoning is that the probabilistic coherence of a set of beliefs and experiences is the same as that of a similar set in which however all the experiences have been replaced by corresponding beliefs. He argues that the first set, the one including experiences, should be accorded a higher degree of justification than the second, which lacks experiences and is nothing but a collection of beliefs.

Berker's idea seems to hinge on a Humean view in which experiences outweigh beliefs. More importantly in the present context, it only bears on models in which probabilistic coherence is a sufficient determinant of justification. For models like ours, in which probabilistic coherence is only necessary, it is not apposite. And of course the phenomenon of fading foundations is not restricted to propositions or beliefs: it manifests itself also in the domain of experiences.

Just as the ground's share in the epistemic justification lessens, so the measure of the ground's causal influence vanishes in the end. In general, whether a regress is epistemic or causal, or whether it is in one or in many dimensions, justification and causation will progressively emerge and foundations will gradually fade away.

Open Access This chapter is licensed under the terms of the Creative Commons Attribution 4.0 International License (http://creativecommons.org/licenses/by/4.0/), which permits use, sharing, adaptation, distribution and reproduction in any medium or format, as long as you give appropriate credit to the original author(s) and the source, provide a link to the Creative Commons license and indicate if changes were made.

The images or other third party material in this chapter are included in the chapter's Creative Commons license, unless indicated otherwise in a credit line to the material. If material is not included in the chapter's Creative Commons license and your intended use is not permitted by statutory regulation or exceeds the permitted use, you will need to obtain permission directly from the copyright holder.

Appendix A
The Rule of Total Probability

Many of the results we use involve an iteration of the rule of total probability: in A.1 we explain how this works in detail. A finite number of iterations leads to a finite regress of probabilities, and in A.2 it is shown how to calculate the maximum error that one can make by limiting oneself to a finite regress. The infinite regress is considered in A.3–A.6; here convergence is demonstrated and the distinction between the usual and the exceptional classes is defined. Attention shifts in A.7 to the peculiar form of a regress of entailments; and finally the Markov condition is put under the theoretical microscope in A.8.

The basic object of interest is a regress of propositions

$$A_0, A_1, A_2, \ldots, A_m, A_{m+1},$$

in which each proposition, except the target A_0, probabilistically supports its neighbour to the left. We first obtain upper and lower bounds on $P(A_n)$, as estimated from the finite chain; and then we look at the infinite regress and show two things:

(i) The infinite series of conditional probabilities is convergent.
(ii) On condition that a certain asymptotic condition is satisfied by the conditional probabilities, the functional dependence of $P(A_n)$ on the value of $P(A_{m+1})$ disappears in the limit $m \to \infty$.

The asymptotic condition will be given explicitly in Section A.4; and the 'usual class' is defined to be the set of all chains of propositions for which this asymptotic condition holds. All other chains belong to the 'exceptional class'. 'Fading foundations' is the name we have given to the phenomenon (ii); and under these conditions $P(A_n)$ is equal to the sum of an infinite series of terms involving conditional probabilities only. In the bulk of the book the interest is in the target proposition, A_0, which is often denoted by q; but in the interests of generality we shall first give formulae for an arbitrary A_n before specializing to the case of the target proposition.

Let us start by recalling some formalism. The unconditional probabilities $P(A_n)$ and $P(A_{n+1})$ are related by the rule of total probability,

$$P(A_n) = \beta_n + \gamma_n P(A_{n+1}), \tag{A.1}$$

with the abbreviations

$$\alpha_n = P(A_n|A_{n+1}) \qquad \beta_n = P(A_n|\neg A_{n+1}) \qquad \gamma_n = \alpha_n - \beta_n. \tag{A.2}$$

The condition of probabilistic support, $\gamma_n > 0$, will be imposed.

A.1 Iterating the rule of total probability

On iterating (A.1) once we obtain

$$\begin{aligned} P(A_n) &= \beta_n + \gamma_n [\beta_{n+1} + \gamma_{n+1} P(A_{n+2})] \\ &= \beta_n + \gamma_n \beta_{n+1} + \gamma_n \gamma_{n+1} P(A_{n+2}). \end{aligned} \tag{A.3}$$

We shall now show that, on iterating (A.1) $m - n$ times, we obtain

$$P(A_n) = \Delta_{n,m} + \Gamma_{n,m} P(A_{m+1}), \tag{A.4}$$

where $\Gamma_{n,m}$ is the finite product

$$\Gamma_{n,m} = \gamma_n \gamma_{n+1} \cdots \gamma_m, \tag{A.5}$$

with $n \leq m$, and where $\Delta_{n,m}$ is the finite sum

$$\Delta_{n,m} = \beta_n + \Gamma_{n,n} \beta_{n+1} + \Gamma_{n,n+1} \beta_{n+2} + \ldots + \Gamma_{n,m-1} \beta_m, \tag{A.6}$$

with $n < m$. Note that $\Delta_{n,m}$ and $\Gamma_{n,m}$ involve conditional probabilities only.

Eq.(A.4) will be proved by the method of mathematical induction. We need to show that, for a fixed n less than m, if (A.4) is true for some particular m, then it is necessarily true with m replaced by $m + 1$. Substitute $m + 1$ for n in Eq.(A.1):

$$P(A_{m+1}) = \beta_{m+1} + \gamma_{m+1} P(A_{m+2}), \tag{A.7}$$

and insert this into Eq.(A.4). The result is

A.2 Extrema of the finite series

$$P(A_n) = \Delta_{n,m} + \Gamma_{n,m}[\beta_{m+1} + \gamma_{m+1}P(A_{m+2})]$$
$$= \Delta_{n,m+1} + \Gamma_{n,m+1}P(A_{m+2}), \quad (A.8)$$

which is (A.4) with $m+1$ written in place of m. Hence, if (A.4) is true for some particular m larger than n, it true for all m larger than n.

Since

$$\Gamma_{n,n} = \gamma_n \; ; \; \Gamma_{n,n+1} = \gamma_n\gamma_{n+1} \; ; \; \Delta_{n,n+1} = \beta_n + \Gamma_{n,n}\beta_{n+1} = \beta_n + \gamma_n\beta_{n+1},$$

it follows that (A.4) is true when $m = n+1$, for (A.4) reduces then to (A.3). In this way the induction has been completed; and (A.4) has been proved to be valid for all $m > n$.

In the special case $n = 0$, (A.4) can be written

$$P(A_0) = \beta_0 + \gamma_0\beta_1 + \gamma_0\gamma_1\beta_2 + \ldots + \gamma_0\gamma_1\ldots\gamma_{n-1}\beta_n + \gamma_0\gamma_1\ldots\gamma_n P(A_{m+1}),$$

which is Eq.(3.20) in Chapter 3.

A.2 Extrema of the finite series

In this section we limit our attention to a finite regress and calculate the maximum and minimum values that the target probability could have, whatever the unknown unconditional probability $P(A_{m+1})$ might be. An approximate value of the target probability is the average of these values; and an upper bound on the error is one half of the difference between the maximum and minimum values. These results are crucial to our discussion of probabilistic justification as a trade-off in Section 5.3.

Thanks to probabilistic support, $\gamma_n > 0$, both $\Delta_{n,m}$ and $\Gamma_{n,m}$ are non-negative. So the minimum value that $P(A_n)$ can have is obtained by setting $P(A_{m+1}) = 0$ in Eq.(A.8); and the maximum value is obtained by setting $P(A_{m+1}) = 1$. Accordingly, $P(A_n)$ is not less than $P_m^{\min}(A_n)$ and not greater than $P_m^{\max}(A_n)$, where

$$P_m^{\min}(A_n) = \Delta_{n,m} \quad (A.9)$$
$$P_m^{\max}(A_n) = \Delta_{n,m} + \Gamma_{n,m}. \quad (A.10)$$

Since

$$P_{m+1}^{\min}(A_n) - P_m^{\min}(A_n) = \Delta_{n,m+1} - \Delta_{n,m} = \Gamma_{n,m}\beta_{m+1} \geq 0,$$

it follows that $P_m^{\min}(A_n)$ is a monotonically increasing function of m. On the other hand,

$$P_{m+1}^{\max}(A_n) - P_m^{\max}(A_n) = \Delta_{n,m+1} - \Delta_{n,m} + \Gamma_{n,m+1} - \Gamma_{n,m} = -\Gamma_{n,m}(1 - \alpha_{m+1}) \leq 0,$$

so $P_m^{\max}(A_n)$ is a monotonically decreasing function of m. This means that the margin of error that one makes by truncating the regress decreases as one adds more links.

In the particular case $n = 0$ we have $P(A_0) \geq \Delta_{0,m}$ and $P(A_0) \leq \Delta_{0,m} + \Gamma_{0,m}$, and these inequalities lead to the following estimates for the target probability and the maximal error committed:

$$P(A_0) = \Delta_{0,m} + \tfrac{1}{2}\Gamma_{0,m} \pm \tfrac{1}{2}\Gamma_{0,m}$$
$$= \beta_0 + \gamma_0\beta_1 + \ldots + \gamma_0\gamma_1 \ldots \gamma_{n-2}\beta_{n-1} + \tfrac{1}{2}\gamma_0\gamma_1 \ldots \gamma_{n-1}\alpha_n \pm \tfrac{1}{2}\gamma_0\gamma_1 \ldots \gamma_n.$$

A.3 Convergence of the infinite series

From (A.2) we have that $\beta_n = \alpha_n - \gamma_n \leq 1 - \gamma_n$, because α_n, being a probability, cannot be greater than unity. Since γ_n is positive, so is $\Gamma_{n,m} = \gamma_n\gamma_{n+1} \ldots \gamma_m$, from which it follows that

$$\Gamma_{n,m}\beta_{m+1} \leq \gamma_n\gamma_{n+1} \ldots \gamma_m(1 - \gamma_{m+1}) = \Gamma_{n,m} - \Gamma_{n,m+1}.$$

Therefore, from (A.6),

$$\Delta_{n,m} \leq \beta_n + (\Gamma_{n,n} - \Gamma_{n,n+1}) + (\Gamma_{n,n+1} - \Gamma_{n,n+2}) + \ldots + (\Gamma_{n,m} - \Gamma_{n,m+1})$$
$$= \beta_n + \Gamma_{n,n} - \Gamma_{n,m+1} \leq \beta_n + \Gamma_{n,n} = \beta_n + (\alpha_n - \beta_n) = \alpha_n.$$

Now $\Delta_{n,m}$ is monotonically increasing as m increases, and since these numbers are bounded by an m-independent number — namely α_n — it follows from the monotone convergence theorem that $\Delta_{n,m}$ has a limit as m tends to infinity. This means that the series

$$P(A_n) = \Delta_{n,\infty} = \beta_n + \Gamma_{n,n}\beta_{n+1} + \Gamma_{n,n+1}\beta_{n+2} + \ldots$$

is convergent.

This proof makes use of the condition of probabilistic support, namely $\gamma_n > 0$ for all n. Convergence (in the usual class) can also be demonstrated without probabilistic support, but since we are interested in epistemic justification, which has probabilistic support as a necessary condition, there is no point in giving the proof without the constraint. Moreover the condition of probabilistic support is essential for justification as a trade-off, see Section 5.3 and Appendix A.2, and for convergence in the probabilistic networks that we discuss in Section 8.4. Incidentally, the condition is also required for convergence in the exceptional class.

In the special case $n = 0$, we conclude that the series (3.24), namely

$$\beta_0 + \gamma_0\beta_1 + \gamma_0\gamma_1\beta_2 + \gamma_0\gamma_1\gamma_2\beta_3 + \cdots,$$

is convergent. In the usual class the series equals the probability of the target proposition.

A.4 When does the remainder term vanish?

We now wish to discover the condition under which the influence of $P(A_{m+1})$ on the value of $P(A_n)$ tends to zero in the limit $m \to \infty$.

Consider first the uniform case, in which the conditional probabilities are the same from link to link. If $\gamma_n = \gamma$, independently of n, then

$$\Gamma_{n,m} = \gamma_n\gamma_{n+1}\cdots\gamma_m = \gamma^{m-n+1}. \tag{A.11}$$

The only exceptional case here is when $\beta_n = 0$ and $\alpha_n = 1$, which corresponds to bi-implication. Apart from this extreme situation, it is the case that $\gamma < 1$, so γ^{m-n+1} tends to zero as m tends to infinity, and therefore $\Gamma_{n,m}$ goes to zero in the infinite limit. As a result the remainder term $\Gamma_{n,m}P(A_{m+1})$ in Eq.(A.4) vanishes too, given that $P(A_{m+1})$ cannot exceed unity.

This result, that $\Gamma_{n,m}$ goes to zero in the limit, generally holds even when the conditional probabilities differ from link to link. For example, if there is a constant, c, less than unity, such that $\gamma_n \leq c < 1$ for all n, then $\Gamma_{n,m} \leq c^{m-n+1}$, which also dies out in the limit. Moreover, this conclusion is usually true even when there is no such constant, c, and γ_n tends to unity. Indeed, $\Gamma_{n,m}$ will be zero in the limit unless γ_m tends very quickly to unity as m goes to infinity — such cases belong to the exceptional class.

To find a precise condition under which the remainder term is equal to zero in the limit, observe that $\gamma_n = \exp(\log \gamma_n) = \exp(-|\log \gamma_n|)$, so

$$\Gamma_{n,\infty} = \gamma_n\gamma_{n+1}\gamma_{n+2}\gamma_{n+3}\cdots = \exp\left[-\sum_{i=n}^{\infty}|\log \gamma_i|\right]. \tag{A.12}$$

Thus $\Gamma_{n,\infty}$ is zero if, and only if, the series $\sum_{i=n}^{\infty}|\log \gamma_i|$ diverges. Since all the terms in this series are positive, the series can only converge, or diverge to

$+\infty$ (it cannot oscillate). If there is a real number $a > 0$, and an integer $N > a$, such that

$$1 - \gamma_n > \frac{a}{n} \quad (A.13)$$

for all $n > N$, then $|\log \gamma_n| > \frac{a}{n}$, and the series diverges, which means that $\Gamma_{n,\infty}$ is zero. Under condition (A.13) the remainder term disappears.

Summarizing, the remainder term generally goes to zero in the limit of infinite m; only in the exceptional class does it fail to do so.

A.5 Example in the usual class

The model

$$\beta_n = \frac{1}{n+3} \qquad \gamma_n = \frac{n+1}{n+2} = 1 - \frac{1}{n+2}, \quad (A.14)$$

belongs to the usual class, since β_n behaves like $1/n$ as n tends to infinity. We find, using the notation of A.1,

$$\Gamma_{n,m} = \gamma_n \gamma_{n+1} \cdots \gamma_m = \tfrac{n+1}{n+2} \times \tfrac{n+2}{n+3} \times \cdots \times \tfrac{m}{m+1} \times \tfrac{m+1}{m+2} = \tfrac{n+1}{m+2}$$

$$\Gamma_{n,m}\beta_{m+1} = \tfrac{n+1}{m+2} \times \tfrac{1}{m+4} = \tfrac{n+1}{2}\left(\tfrac{1}{m+2} - \tfrac{1}{m+4}\right).$$

Hence

$$\Delta_{n,m} = \beta_n + \Gamma_{n,n}\beta_{n+1} + \Gamma_{n,n+1}\beta_{n+2} + \ldots + \Gamma_{n,m-1}\beta_m$$

$$= \tfrac{1}{n+3} + \tfrac{n+1}{2}\left[\left(\tfrac{1}{n+2} - \tfrac{1}{n+4}\right) + \left(\tfrac{1}{n+3} - \tfrac{1}{m+5}\right) + \left(\tfrac{1}{n+4} - \tfrac{1}{m+6}\right) + \ldots\right.$$

$$\left. \ldots + \left(\tfrac{1}{m-1} - \tfrac{1}{m+1}\right) + \left(\tfrac{1}{m} - \tfrac{1}{m+2}\right) + \left(\tfrac{1}{m+1} - \tfrac{1}{m+3}\right)\right]$$

$$= \tfrac{1}{n+3} + \tfrac{n+1}{2}\left(\tfrac{1}{n+2} + \tfrac{1}{n+3} - \tfrac{1}{m+2} - \tfrac{1}{m+3}\right) = 1 - \tfrac{1}{2}\tfrac{1}{n+2} - \tfrac{1}{2}\tfrac{(n+1)(2m+5)}{(m+2)(m+3)}.$$

From Eq.(A.4) we have then

$$P(A_n) = 1 - \frac{1}{2(n+2)} - \frac{(n+1)(2m+5)}{2(m+2)(m+3)} + \frac{n+1}{m+2}P(A_{m+1}). \quad (A.15)$$

In the limit of an infinite linear chain, $m \to \infty$, we obtain

$$P(A_n) = 1 - \frac{1}{2(n+2)}.$$

In the particular case $n = 0$, Eq.(A.15) becomes

A.6 Example in the exceptional class

$$P(A_0) = \frac{3}{4} - \frac{2m+5}{2(m+2)(m+3)} + \frac{1}{m+2}P(A_{m+1}), \quad (A.16)$$

which is Eq.(3.22) in Chapter 3.

On the other hand, if the A_n form a finite loop instead of an infinite chain, with $A_{m+1} = A_0$, then (A.16) reads

$$P(A_0) = \frac{3}{4} - \frac{2m+5}{2(m+2)(m+3)} + \frac{1}{m+2}P(A_0); \quad (A.17)$$

and this can be solved for $P(A_0)$ to yield

$$P(A_0) = \frac{3}{4} - \frac{1}{4(m+3)}, \quad (A.18)$$

which is Eq.(8.3). On substituting the value (A.18) into (A.15) — with $P(A_0)$ in place of $P(A_{m+1})$ — we finally obtain the probability at an arbitrary site on the loop:

$$P(A_n) = 1 - \frac{1}{2(n+2)} - \frac{n+1}{4(m+3)},$$

which is valid for $0 \leq n \leq m$.

A.6 Example in the exceptional class

An example in the exceptional class is

$$\beta_n = \frac{1}{(n+2)(n+3)} \qquad \gamma_n = \frac{(n+1)(n+3)}{(n+2)^2} = 1 - \frac{1}{(n+2)^2}.$$

The crucial difference is that here β_n and $1 - \alpha_n = 1 - \beta_n - \gamma_n$ both tend to zero as fast as $1/n^2$, as n tends to infinity. To derive $P(A_n)$ we first calculate

$$\Gamma_{n,m} = \gamma_n \ldots \gamma_m = \left[\frac{n+1}{n+2} \cdot \frac{n+3}{n+2}\right] \times \ldots \times \left[\frac{m+1}{m+2} \cdot \frac{m+3}{m+2}\right] = \frac{n+1}{n+2} \cdot \frac{m+3}{m+2}$$

$$\Gamma_{n,m}\beta_{m+1} = \frac{n+1}{n+2}\frac{m+3}{m+2} \times \frac{1}{(m+3)(m+4)} = \frac{1}{2}\frac{n+1}{n+2}\left(\frac{1}{m+2} - \frac{1}{m+4}\right).$$

After some algebra we find

$$\Delta_{n,m} = \beta_n + \Gamma_{n,n}\beta_{n+1} + \Gamma_{n,n+1}\beta_{n+2} + \ldots + \Gamma_{n,m-1}\beta_m$$
$$= \frac{1}{2}\frac{2n+3}{(n+2)^2} - \frac{1}{2}\frac{n+1}{n+2}\frac{2m+5}{(m+2)(m+3)}.$$

From Eq.(A.4) we deduce that[1]

---
[1] An easier way to derive this equation is by putting $Q_n = \frac{n+2}{n+1}P(A_n)$ in Eq.(A.1), which leads to

$$P(A_n) = \frac{1}{2}\frac{2n+3}{(n+2)^2} - \frac{1}{2}\frac{n+1}{n+2}\cdot\frac{2m+5}{(m+2)(m+3)} + \frac{n+1}{n+2}\cdot\frac{m+3}{m+2}P(A_{m+1}).$$

In the particular case $n = 0$, this becomes

$$P(A_0) = \frac{3}{8} - \frac{2m+5}{4(m+2)(m+3)} + \frac{1}{2}\frac{m+3}{m+2}P(A_{m+1}), \quad (A.19)$$

which is Eq.(3.26). In the limit that m tends to infinity we find formally

$$P(A_0) = \frac{3}{8} + \frac{1}{2}P(A_\infty),$$

where $P(A_\infty)$ is an indeterminate number in the interval $[0, 1]$. However, for the infinite loop we can set $P(A_{m+1}) = P(A_0)$ in (A.19) and solve the linear equation for $P(A_0)$:

$$P(A_0) = \left[\frac{3}{8} - \frac{2m+5}{4(m+2)(m+3)}\right] \bigg/ \left[1 - \frac{1}{2}\frac{m+3}{m+2}\right]$$

$$= \frac{3}{4} - \frac{1}{4(m+3)},$$

which is Eq.(8.6).

A.7 The regress of entailment

The classical regress is one of entailment, in which every proposition, A_{n+1}, entails the proposition to its left, A_n, for all $n = 0, 1, 2, \ldots$. In this case, $\alpha_n = P(A_n|A_{n+1}) = 1$ for all n. From (A.2) we have that

$$\beta_n = \alpha_n - \gamma_n = 1 - \gamma_n.$$

Then Eq.(A.6) takes on the form

$$\Delta_{n,m} = 1 - \gamma_n + \Gamma_{n,n}(1-\gamma_{n+1}) + \Gamma_{n,n+1}(1-\gamma_{n+2}) + \ldots + \Gamma_{n,m-1}(1-\gamma_m)$$
$$= 1 - \Gamma_{n,m}. \quad (A.20)$$

$$Q_n = \frac{1}{2}\left(\frac{1}{n+1} - \frac{1}{n+3}\right) + Q_{n+1}.$$

It is a simple matter to concatenate this relation to obtain the relation between Q_n and Q_{m+1}, and thence that between $P(A_n)$ and $P(A_{m+1})$. A similar ploy could have been used in Section A.5 by means of the substitution $Q_n = \frac{1}{n+1}P(A_n)$; but in this case not much labour would have been saved.

From Eq.(A.4) we then obtain

$$P(A_n) = 1 - \Gamma_{n,m} + \Gamma_{n,m} P(A_{m+1}),$$

which is equivalent to

$$P(\neg A_n) = \Gamma_{n,m} P(\neg A_{m+1}). \tag{A.21}$$

In the special case $n = 0$, this reads

$$P(\neg A_0) = \gamma_0 \gamma_1 \ldots \gamma_m P(\neg A_{m+1}),$$

which is Eq.(3.27).

A.8 Markov condition and conjunctions

In our discussion of causal chains in Section 8.6, we remarked that a way of demonstrating that there indeed exists a probability distribution over all the possible conjunctions of the propositions in a probabilistic regress was to impose a suitable Markov condition. Here we show how to construct the probability of a typical conjunction.

Suppose then that the following Markov condition holds:

$$P(A_n | \pm A_{n+1} \wedge Z) = P(A_n | \pm A_{n+1}), \tag{A.22}$$

for all n, where $\pm A_{n+1}$ means A_{n+1} or $\neg A_{n+1}$, and where Z stands for any event, A_m, such that $m \geq n+2$, or its negation, or for any conjunction of such events, or their negations. We shall illustrate how one can calculate the probability of any conjunction of the A_n, or their negations, by working out one representative example in detail:

$$\begin{aligned}
P(A_1 \wedge \neg A_3 \wedge A_4) &= P(A_1 \wedge A_2 \wedge \neg A_3 \wedge A_4) + P(A_1 \wedge \neg A_2 \wedge \neg A_3 \wedge A_4) \\
&= P(A_1 | A_2 \wedge \neg A_3 \wedge A_4) P(A_2 \wedge \neg A_3 \wedge A_4) \\
&\quad + P(A_1 | \neg A_2 \wedge \neg A_3 \wedge A_4) P(\neg A_2 \wedge \neg A_3 \wedge A_4) \\
&= P(A_1 | A_2) P(A_2 | \neg A_3 \wedge A_4) P(\neg A_3 \wedge A_4) \\
&\quad + P(A_1 | \neg A_2) P(\neg A_2 | \neg A_3 \wedge A_4) P(\neg A_3 \wedge A_4) \\
&= P(A_1 | A_2) P(A_2 | \neg A_3) P(\neg A_3 | A_4) P(A_4) \\
&\quad + P(A_1 | \neg A_2) P(\neg A_2 | \neg A_3) P(\neg A_3 | A_4) P(A_4) \\
&= [\alpha_1 \beta_2 + \beta_1 (1 - \beta_2)](1 - \alpha_3) P(A_4) \\
&= (\beta_1 + \gamma_1 \beta_2)(1 - \alpha_3) P(A_4).
\end{aligned}$$

The unconditional probability is given by the following convergent series of terms that only involve the conditional probabilities:

$$P(A_4) = \beta_4 + \gamma_4 \beta_5 + \gamma_4 \gamma_5 \beta_6 + \gamma_4 \gamma_5 \gamma_6 \beta_7 + \ldots.$$

It is assumed that the set of conditional probabilities belongs to the usual class. Any other conjunction of propositions or events A_n and $\neg A_m$ can be handled in an analogous manner.

An interesting consequence of the imposition of the Markov condition has to do with the possible transitivity of probabilistic support. To see this, consider the following measure of the probabilistic support that A_m gives to A_n:

$$S(A_n, A_m) = P(A_n|A_m) - P(A_n|\neg A_m).$$

According to (A.2), $S(A_n, A_{n+1}) = \gamma_n$. We shall show that, under the Markov condition, and for any m larger than n, $S(A_n, A_m) = \gamma_n \gamma_{n+1} \ldots \gamma_{m-1}$. The condition of probabilistic support means that all the γ_n are positive, and so $S(A_n, A_m)$ is also positive for all $n < m$. This shows that probabilistic support is transitive under the Markov condition.[2] Although the ground, A_{m+1}, supports the target, A_0, it does so to a degree that becomes smaller and smaller as the chain gets longer and longer. In the usual class the product $\gamma_0 \gamma_1 \gamma_2 \ldots$ diverges to zero, so the support that A_{m+1} gives to A_0 dwindles away to nothing as m tends to infinity, whereas the exceptional class it is positive, so in this case the support, although it continues to dwindle, does not go all the way to zero.[3]

To prove that, under the Markov condition,

$$S(A_n, A_m) = \gamma_n \gamma_{n+1} \ldots \gamma_{m-1}, \qquad (A.23)$$

we take recourse again to the method of mathematical induction. Consider

$$P(A_n \wedge A_{m+1}) = P(A_n \wedge A_m \wedge A_{m+1}) + P(A_n \wedge \neg A_m \wedge A_{m+1})$$
$$= P(A_n | A_m \wedge A_{m+1}) P(A_m \wedge A_{m+1}) + P(A_n | \neg A_m \wedge A_{m+1}) P(\neg A_m \wedge A_{m+1})$$
$$= P(A_n | A_m) P(A_m | A_{m+1}) P(A_{m+1}) + P(A_n | \neg A_m) P(\neg A_m | A_{m+1}) P(A_{m+1}),$$

where the last line follows because of the Markov condition. On dividing by $P(A_{m+1})$ we find

---

[2] The first proof of the transitivity of probabilistic support under a condition of screening off is in Reichenbach 1956 on page 160, Eq.(12). A later proof can be found in Shogenji 2003.

[3] This dwindling of support as the chain increases in length was noted in Roche and Shogenji 2014.

A.8 Markov condition and conjunctions

$$P(A_n|A_{m+1}) = P(A_n|A_m)P(A_m|A_{m+1}) + P(A_n|\neg A_m)P(\neg A_m|A_{m+1}). \quad \text{(A.24)}$$

Similarly, replacing A_{m+1} by $\neg A_{m+1}$ in the above reasoning, we obtain

$$P(A_n|\neg A_{m+1}) = P(A_n|A_m)P(A_m|\neg A_{m+1}) + P(A_n|\neg A_m)P(\neg A_m|\neg A_{m+1}). \quad \text{(A.25)}$$

Subtracting Eq.(A.25) from Eq.(A.24), we see that

$$\begin{aligned} P(A_n|A_{m+1}) - P(A_n|\neg A_{m+1}) &= P(A_n|A_m)[P(A_m|A_{m+1}) - P(A_m|\neg A_{m+1})] \\ &\quad + P(A_n|\neg A_m)[P(\neg A_m|A_{m+1}) - P(\neg A_m|\neg A_{m+1})] \\ &= [P(A_n|A_m) - P(A_n|\neg A_m)][P(A_m|A_{m+1}) - P(A_m|\neg A_{m+1})]. \end{aligned}$$

That is, under the Markov condition,

$$S(A_n, A_{m+1}) = S(A_n, A_m)S(A_m, A_{m+1}).$$

Now $S(A_m, A_{m+1}) = \gamma_{m-1}$, so if Eq.(A.23) is true for some $m > n$, then

$$S(A_n, A_{m+1}) = \gamma_n \gamma_{n+1} \cdots \gamma_{m-1} \gamma_m, \quad \text{(A.26)}$$

which has the same form as (A.23), with m replaced by $m+1$. Since (A.23) is true for $m = n+1$, the induction is complete.

Appendix B
Closure Under Conjunction

In Section 6.5 we noted that Tomoji Shogenji has constructed a measure of justification that takes account of intuitions regarding closure and independence. Here we shall spell out this measure, J, by the method of one of us.[1] If $J(h,e)$ is a continuous function of $x = P(h|e)$ and $y = P(h)$ only, we may write

$$J(h,e) = F(x,y), \tag{B.1}$$

where $F(x,y)$ is a continuous function for $x \in [0,1]$ and $y \in (0,1)$. Discontinuities or divergences are allowed if $P(h)$ is extremal (0 or 1), but continuity with respect to the conditional probability, $P(h|e)$, is required at both end points.

Let h_1, h_2 and e be propositions such that

$$P(h_1|e) = P(h_2|e) = x$$
$$P(h_1) = P(h_2) = y$$

and let h_1 and h_2 be independent of one another, conditionally with respect to e, and also unconditionally:

$$P(h_1 \wedge h_2|e) = P(h_1|e)P(h_2|e) = x^2$$
$$P(h_1 \wedge h_2) = P(h_1)P(h_2) = y^2.$$

If $J(h_1, e) = s$ and $J(h_2, e) = s$, then it is required that also $J(h_1 \wedge h_2, e) = s$. Thus $J(h_1 \wedge h_2, e) = J(h_1, e)$, and so, from Eq.(B.1),

$$F(x,y) = F(x^2, y^2). \tag{B.2}$$

Change the variables and the function from $F(x,y)$ to $G(x,u)$, where

---
[1] Shogenji 2012; Atkinson 2012.

$$u = \frac{\log x}{\log y} \qquad G(x,u) = F(x,y).$$

Condition (B.2) becomes

$$G(x,u) = G(x^2,u).$$

For any $x \in (0,1)$, we can iterate this equation to obtain

$$G(x,u) = G(x^2,u) = G(x^4,u) = \ldots = G(x^{2^n},u).$$

Since the function $G(x,u)$ is required to be continuous at $x = 0$, we can take the limit $n \to \infty$ and conclude that $G(x,u) = G(0,u) \equiv f(u)$ is an arbitrary continuous function of u. Hence

$$J(h,e) = f\left(\frac{\log P(h|e)}{\log P(h)}\right). \tag{B.3}$$

$J(h,e)$ is an increasing function of $P(h|e)$ and a decreasing function of $P(h)$, so it follows that $f(u)$ must be a decreasing function of u (since $\log P(h|e)$ and $\log P(h)$ are both negative). The most general function of justification that satisfies Eq.(B.2) has the form (B.3), subject to the constraint that $f(u)$ is a continuous, monotonically decreasing function of u.

We shall generalize this result by supposing now only that $J(h_1,e) \geq s$ and $J(h_2,e) \geq s$, instead of the more restrictive $J(h_1,e) = s$ and $J(h_2,e) = s$. So

$$\frac{\log P(h_1|e)}{\log P(h_1)} \leq f^{-1}(s) \quad \text{and} \quad \frac{\log P(h_2|e)}{\log P(h_2)} \leq f^{-1}(s),$$

where the inverse function, f^{-1}, is guaranteed to exist, given the monotonicity of f. Then

$$\begin{aligned}
\log[P(h_1 \wedge h_2|e)] &= \log[P(h_1|e)P(h_2|e)] \\
&= \log P(h_1|e) + \log P(h_2|e) \\
&\geq f^{-1}(s)[\log P(h_1) + \log P(h_2)] \\
&= f^{-1}(s) \log[P(h_1)P(h_2)] \\
&= f^{-1}(s) \log[P(h_1 \wedge h_2)].
\end{aligned}$$

Therefore, remembering again that the logarithms are negative, we have

$$\frac{\log P(h_1 \wedge h_2|e)}{\log P(h_1 \wedge h_2)} \leq f^{-1}(s),$$

B Closure Under Conjunction

and so $J(h_1 \wedge h_2, e) \geq s$. A similar proof works with the inequalities working in the opposite direction, i.e. if $J(h_1, e) \leq s$ and $J(h_2, e) \leq s$ then $J(h_1 \wedge h_2, e) \leq s$. Moreover, the method extends straightforwardly to an arbitrary finite number of independent hypotheses $h_1, h_2, \ldots, h_n$, instead of two. This concludes the demonstration that Eq.(B.3) encapsulates the most general measure of justification.

All measures that satisfy the above conditions are ordinally equivalent to one another. For consider two different measures:

$$J_1(h,e) = f_1\left[\frac{\log P(h|e)}{\log P(h)}\right] \quad \text{and} \quad J_2(h,e) = f_2\left[\frac{\log P(h|e)}{\log P(h)}\right].$$

Because f_1 is a monotonically decreasing function, a necessary and sufficient condition that $J_1(h_1, e_1) > J_1(h_2, e_2)$, is

$$\frac{\log P(h_1|e_1)}{\log P(h_1)} < \frac{\log P(h_2|e_2)}{\log P(h_2)},$$

and because of the monotonicity of f_2, this is a necessary and sufficient condition that $J_2(h_1, e_1) > J_2(h_2, e_2)$. Analogous reasoning holds if the sign $>$ is replaced by $<$ or by $=$. Thus all measures of justification are ordinally equivalent to one another.

If h and e are such that $P(h|e) = P(h)$, then $J(h,e) = f(1)$, irrespective of the value of $P(h) \in (0,1)$. This is the condition of equineutrality, and we conventionally set $f(1) = 0$. If, on the other hand, h and e are such that $P(h|e) = 1$, then $J(h,e) = f(0)$, irrespective of the value of $P(h) \in (0,1)$. This is the condition of equimaximality, and we set $f(0) = 1$.

The simplest realization of the above constraints is $f(u) = 1 - u$, which leads to

$$J(h,e) = 1 - \frac{\log P(h|e)}{\log P(h)} = \frac{\log P(h|e) - \log P(h)}{-\log P(h)}. \tag{B.4}$$

If $J(h,e) \geq s$, then

$$\frac{\log P(h|e)}{\log P(h)} \leq 1 - s,$$

and, since $\log P(h)$ is negative, it follows that

$$\log P(h|e) \geq (1-s)\log P(h) = \log[P(h)]^{1-s},$$

which entails

$$P(h|e) \geq [P(h)]^{1-s}.$$

With q in place of h and A_1 in place of e, this reads

$$P(q|A_1) \geq [P(q)]^{1-s},$$

which is the inequality (6.12) of Chapter 6.

Appendix C
Washing Out of the Prior

There is a much-vaunted escape clause that Bayesians use when they are confronted with an unsatisfactory feature of their method. The unsatisfactory feature is that the final, or posterior probability of a hypothesis depends on its prior probability, which is to a large extent arbitrary. The escape clause is that repeated updatings of the same hypothesis by more and more evidence lead, in favourable circumstances, to the 'washing out' of the prior, i.e. the insensitivity of the final posterior probability to the precise value that the prior probability might have. In the formally infinite limit the posterior is independent of the prior.

A probabilistic regress, within the usual class, has a superficially similar property that we have dubbed 'fading foundations'. The probability of the target depends less and less on the probability of the ground as the chain of propositions becomes longer and longer, and in the formal limit of an infinite regress it is independent of the ground.

In the next section Bayesian washing out is explained in intuitive terms, and then an example is given concerning the bias of a bent coin. In section C.3 we point out in detail why Bayesian washing out is quite different from fading foundations.

C.1 Washing out

Suppose we have some evidence, e_1, for a hypothesis, p_1, and we can calculate the likelihood with which e_1 would obtain if hypothesis p_1 were true, namely $P(e_1|p_1)$. What we want is rather the probability that p_1 is correct, given that e_1 is true, and this we calculate from Bayes' formula:

$$P(p_1|e_1) = \frac{P(e_1|p_1)P_0(p_1)}{P(e_1)}. \tag{C.1}$$

Here $P_0(p_1)$ is the *prior* probability that is accorded to the hypothesis p_1: some Bayesians allow this to depend wholly on whim, others require it to be determined by some previous knowledge of the situation in question. In any case $P_0(p_1)$ is to be superseded by the posterior probability, or update, $P_1(p_1) = P(p_1|e_1)$. The denominator in (C.1) can be computed from the rule of total probability:

$$P(e_1) = P(e_1|p_1)P_0(p_1) + P(e_1|\neg p_1)[1 - P_0(p_1)], \tag{C.2}$$

on condition that the likelihood $P(e_1|\neg p_1)$ can also be calculated. More generally, if $\{p_i\}$, $i = 1, 2, \ldots, n$, is a partition of the space of hypotheses for the situation in question, i.e. $p_i \wedge p_j$ is impossible for all $i \neq j$, and the disjunction $p_1 \vee p_2 \vee \ldots \vee p_n$ is the whole space, then (C.2) is replaced by

$$P(e_1) = P(e_1|p_1)P_0(p_1) + P(e_1|p_2)P_0(p_2) + \ldots + P(e_1|p_n)P_0(p_n). \tag{C.3}$$

Suppose now that some new evidence, e_2, comes in. The old posterior probability, $P_1(p_1) = P(p_1|e_1)$, serves as the new prior, and the new posterior probability is $P_2(p_1) = P_1(p_1|e_2) = P(p_1|e_1 \wedge e_2)$. After m pieces of new evidence have come in, $P_m(p_1) = P(p_1|e_1 \wedge e_2 \wedge \ldots \wedge e_m)$ is the final posterior probability. The idea is that, if p_1 is the correct hypothesis, and $p_2, p_3, \ldots, p_m$ are all incorrect hypotheses, the likelihood $P(e_1 \wedge e_2 \wedge \ldots \wedge e_m|p_1)$ will become larger and larger as more and more data comes in, that is, as m increases, and all the $P(e_1 \wedge e_2 \wedge \ldots \wedge e_m|p_i)$ with $i \neq 1$ will become smaller and smaller. This means that, for large m, $P(e_1 \wedge e_2 \wedge \ldots \wedge e_m)$ will be equal to $P(e_1 \wedge e_2 \wedge \ldots \wedge e_m|p_1)P_0(p_1)$ in good approximation, since the other terms, depending on $p_2, p_3, \ldots, p_n$, will be negligible. Hence $P(p_1|e_1 \wedge e_2 \wedge \ldots \wedge e_m)$ will be close to 1, and it thus may be expected that

$$P_m(p_1) = P(p_1|e_1 \wedge e_2 \wedge \ldots \wedge e_m) \tag{C.4}$$

will tend to 1 in the limit. Note that the original prior probability, $P_0(p_1)$, has cancelled, that is to say, it has 'washed out'.

This was a quick and dirty explanation of how repeated Bayesian updatings can be expected to lead one to the true hypothesis as more and more evidence is accumulated. A more sophisticated treatment can be found for example in the *Stanford Encyclopedia of Philosophy* (Hawthorne 2014). We shall now exhibit this effect explicitly in the case of a bent coin that is tossed repeatedly, the purpose being to ascertain its bias.

C.2 Example: a bent coin

Suppose that p is the bias of a bent coin, i.e. the probability that heads will come up when the coin is tossed. Let e_1 stand for the evidence that h_1 heads have turned up in n_1 tosses. The likelihood $P(e_1|p)$, the conditional probability that e_1 would result, is

$$P(e_1|p) = \frac{n_1!}{h_1!(n_1-h_1)!} p^{h_1}(1-p)^{n_1-h_1},$$

the factor involving the factorials being the number of different ways that h_1 heads can turn up in n_1 tosses.

Suppose though that the bias, p, is unknown. We are interested in the inverse conditional probability, $P(p|e_1)$, i.e. the probability that a head will turn up given the evidence e_1. Here is Bayes's theorem again:

$$P(p|e_1) = \frac{P(e_1|p)P_0(p)}{P(e_1)}. \tag{C.5}$$

As before, $P_0(p)$ is the Bayesian prior, a subjective guess which is to be updated by (C.5), on the basis of the evidence, e_1. Strictly speaking, $P_0(p)$ is not a probability, but rather a probability *density:* the prior probability that the bias lies between p and $p+dp$ is the infinitesimal $P_0(p)dp$.

The denominator in (C.5) can be written as a continuous partition of the probability space as follows:

$$P(e_1) = \int_0^1 dp\, P(e_1|p)P_0(p) = \frac{n_1!}{h_1!(n_1-h_1)!} \int_0^1 dp\, p^{h_1}(1-p)^{n_1-h_1} P_0(p). \tag{C.6}$$

This takes the place of the sum (C.3) in the discrete case that was considered in the previous section. Following the exposition of Howson and Urbach (2006), we will insert for the prior a so-called beta distribution:

$$P_0(p) = B(u,v)\, p^{u-1}(1-p)^{v-1}, \tag{C.7}$$

where u and v are to be regarded as free parameters that can be varied to give an idea of the arbitrariness that is inherent in the Bayesian approach, and where $B(u,v)$ is a normalization factor that need not be specified, since it will cancel. The adoption of (C.7) as the prior has no good justification, except the rather lame

> ... beta distributions take on a wide variety of shapes, depending on the values of two parameters, u and v, enabling you to choose a beta distribution that best approximates your actual distribution of beliefs.[1]

---
[1] Howson and Urbach 2006, 242.

On substituting (C.7) into (C.6), we find that we can evaluate the integral. It is in fact a beta-function (and that is the main reason, but of course not a justification, for choosing (C.7) in the first place). The result is

$$P(p|e_1) = \frac{(n_1+u+v-1)!}{(h_1+u-1)!(n_1-h_1+v-1)!} p^{h_1+u-1}(1-p)^{n_1-h_1+v-1}. \quad \text{(C.8)}$$

This is the posterior probability density corresponding to the value p. That is not quite what we were looking for, since it does not give one value for the probability associated with our bent coin, but rather a whole spread of values. But this is as it should be: one single value for the probability is not singled out as the only possibility. We need to calculate the mean value of p according to the distribution (C.8), which will give the most likely value for the sought-for probability, and the standard deviation, which will indicate how uncertain the estimate is.

Straightforward calculations yield

$$\bar{p}_B \equiv E[p] = \frac{h_1+u}{n_1+u+v}$$

$$\sigma_B^2 \equiv E[(p-\bar{p}_B)^2] = \frac{\bar{p}_B \bar{q}_B}{n_1+u+v+1},$$

where $\bar{q}_B = 1 - \bar{p}_B$, and where the subscript 'B' is to remind us that the mean and standard deviation here are Bayesian estimates. The uniform prior, $P_0(p) = 1$, which corresponds to the choice $u = 1 = v$, gives the mean $\bar{p}_B = (h_1+1)/(n_1+2)$, which is the celebrated result of Laplace. Let us however keep u and v general, since the Laplacean choice is merely one of an infinite number of possibilities.

Suppose that a second run of n_2 tosses is made, in which h_2 heads come up, and take the posterior density (C.8) after the first run of n_1 tosses as the prior density for the second run. With e_2 denoting the evidence relating to the latter run, it is evident that the new posterior probability density will be

$$P(p|e_1 \wedge e_2) \propto p^{h_1+h_2+u-1}(1-p)^{n_1+n_2-h_1-h_2+v-1},$$

where we have suppressed the normalization factor. More generally, after many runs, with sequential updating, the posterior probability density is proportional to

$$p^{h+u-1}(1-p)^{n-h+v-1},$$

where n is the total number of tosses in all the runs, and h is the total number of heads that have come up. The mean is $(h+u)/(n+u+v)$, which becomes closer and closer to the relative frequency, h/n, as n increases, the standard deviation becoming smaller and smaller. The prior, specified by the constants u and v, washes out in the limit.

So the success of repeated Bayesian updating lies simply in its tending to the relative frequency. A statistician might well be forgiven for pointing out that one does not need a Bayesian prior and the rigmarole of Bayesian updating to come to the conclusion that the expected value of the ratio of the number of heads to the number of tosses is equal to the bias of the coin.

C.3 Washing out is not fading away

At first sight there might seem to be a similarity between:

1. The washing out of the prior in Bayesian updating, that is the independence in the infinite limit of the posterior on the prior, and
2. The fading of the foundation in a probabilistic regress, that is the independence in the infinite limit of the target probability on the probability of the ground.

However, the two effects are very different. In Bayesian updating the Bayes formula (C.1) involves the computation of $P(p_1|e_1)$ in terms of the inverse conditional probability, $P(e_1|p_1)$, followed by $P(p_1|e_1 \wedge e_2)$, and so on, as more evidence accumulates. This is quite different from our calculation of $P(q)$, in which there is no inversion à la Bayes, but rather a sequence of propositions, $A_1, A_2, \ldots$ that follow one another in a linear chain. In a sense the dissimilarity between the two could not have been greater. Fading foundations implies that the more distant propositions in the chain have less influence on the probability of the target than do the first few propositions. In Bayesian updating, on the contrary, the various pieces of evidence, although they are introduced one after another, are actually all on a par, as can be seen from (C.4).

Appendix D
Fixed-Point Methods

In Section 3.4 we analyzed the one-dimensional uniform chain of propositions by summing a geometrical series. Below, in D.1, we show how a fixed-point method can be used to obtain the same result. This serves as an introduction to the fixed-point analysis in D.2 of the more complicated case of the two-dimensional uniform network that was discussed in Section 8.4.

It is important to note that the fixed-point method, both in the one-dimensional and the many-dimensional cases, only works if there is uniformity, i.e. if the conditional probabilities remain the same throughout the chain or network. The analysis that we gave for the one-dimensional chain in the text is therefore more general, since it also applies if the conditional probabilities are not uniform.

D.1 Linear iteration

In Section 3.4 we considered a recursion relation for a uniform chain of propositions that has the form

$$P(A_n) = \alpha P(A_{n+1}) + \beta P(\neg A_{n+1})$$
$$= \beta + (\alpha - \beta) P(A_{n+1}). \quad \text{(D.1)}$$

Here A_0 is the target proposition, which we sometimes wrote as q. In 3.4 we explained how to calculate $P(q)$ by summing a geometric series.

Here is the same story in terms of fixed points. The question is: is there a special value of $P(A_{n+1})$, say p_*, such that if we plug it into the right-hand side of (D.1), the very same value, p_*, results for $P(A_n)$? Indeed there is, for a unique solution of the equation

$$p_* = \beta + (\alpha - \beta) p_*, \tag{D.2}$$

exists, namely

$$p_* = \frac{\beta}{1 - \alpha + \beta},$$

given that the condition of probabilistic support implies $0 < \alpha - \beta < 1$. This agrees with what we found in Section 3.7 for $P(q)$. We still have to do more work, however, before concluding that p_* is an attracting fixed point of the iteration (D.1); and it will be salutary to sketch what is involved. From (D.1) and (D.2) we see that

$$P(A_n) - p_* = (\alpha - \beta)(P(A_{n+1}) - p_*).$$

Since $\alpha - \beta$ is less than one, it follows that the distance between $P(A_n)$ and p_*, if it is not zero, will be less than the distance between $P(A_{n+1})$ and p_*; and the distance between $P(A_{n-1})$ and p_* will be smaller still. If we start the iteration at a very large value of n, and iterate down to $n = 0$, that is down to the target proposition q, we will find that

$$P(q) - p_* = (\alpha - \beta)^{n+1} (P(A_{n+1}) - p_*).$$

Because $(\alpha - \beta)^{n+1}$ will be very small for large n, it is the case that, whatever value we choose for $P(A_{n+1})$, the difference between $P(q)$ and p_* will be tiny; and, in the limit of infinite n, that is for an infinite chain of bacterial ancestors, $P(q) = p_*$.

Here p_* is the attracting fixed point of the iteration (D.1). One can express the essence of this as follows:

$$p' = \beta + (\alpha - \beta)p,$$

where one starts with some value for p, and then puts the resulting value of p' back into the right-hand side as a new value for p. This procedure is repeated *ad infinitum* in thought. The fixed point p_* attracts p to itself in this process.

D.2 Quadratic Iteration

In Section 8.3 we obtained the recursion relation (8.9), namely

$$P(A_n) = \alpha_n P^2(A_{n+1}) + \beta_n P^2(\neg A_{n+1}) + (\gamma_n + \delta_n) P(A_{n+1}) P(\neg A_{n+1}).$$

D.2 Quadratic Iteration

Using the fact that the conditional probabilities α_n, β_n, γ_n and δ_n are all non-negative, we see from (8.9) that, if $P(A_{n+1})$ lies within the unit interval (so that $P(\neg A_{n+1})$ does so too), then $P(A_n) \geq 0$. Moreover, since α_n, β_n, γ_n and δ_n are each not greater than one,

$$P(A_n) \leq P^2(A_{n+1}) + P^2(\neg A_{n+1}) + 2P(A_{n+1})P(\neg A_{n+1})$$
$$= [P(A_{n+1}) + P(\neg A_{n+1})]^2 = 1. \quad (D.3)$$

Thus we have demonstrated that $0 \leq P(A_{n+1}) \leq 1$ entails $0 \leq P(A_n) \leq 1$, and this means that the quadratic iteration will not run amok: the probabilities remain within the unit interval, as they should; and the question is whether $P(A_0)$ tends to a limit as the length of the chain tends to infinity, or whether it wanders around indefinitely.

When the conditional probabilities do not change from link to link, we may drop the indices; and, with the substitution of $1 - P(A_{n+1})$ for $P(\neg A_{n+1})$ in (8.9), we obtain

$$P(A_n) = \alpha P^2(A_{n+1}) + \beta \left(1 - 2P(A_{n+1}) + P^2(A_{n+1})\right)$$
$$+ (\gamma + \delta)\left(P(A_{n+1}) - P^2(A_{n+1})\right)$$
$$= \beta + 2(\varepsilon - \beta)P(A_{n+1}) + (\alpha + \beta - 2\varepsilon)P^2(A_{n+1}), \quad (D.4)$$

with $\varepsilon = \frac{1}{2}(\gamma + \delta)$. On condition that $\alpha + \beta \neq 2\varepsilon$, define

$$q_n = (\alpha + \beta - 2\varepsilon)P(A_n) - \beta + \varepsilon$$
$$= (\alpha + \beta - 2\varepsilon)[\beta + 2(\varepsilon - \beta)P(A_{n+1}) + (\alpha + \beta - 2\varepsilon)P^2(A_{n+1})] - \beta + \varepsilon$$
$$= \beta(\alpha + \beta - 2\varepsilon) - \beta + \varepsilon +$$
$$2(\varepsilon - \beta)(\alpha + \beta - 2\varepsilon)P(A_{n+1}) + (\alpha + \beta - 2\varepsilon)^2 P^2(A_{n+1}). \quad (D.5)$$

This definition of q_n also implies that

$$q_{n+1}^2 = [(\alpha + \beta - 2\varepsilon)P(A_{n+1}) - \beta + \varepsilon]^2 \quad (D.6)$$
$$= (\varepsilon - \beta)^2 +$$
$$2(\varepsilon - \beta)(\alpha + \beta - 2\varepsilon)P(A_{n+1}) + (\alpha + \beta - 2\varepsilon)^2 P^2(A_{n+1}).$$

Comparing (D.5) with (D.7), we see that $q_n = c + q_{n+1}^2$, where

$$c = \beta(\alpha + \beta - 2\varepsilon) - \beta + \varepsilon - (\varepsilon - \beta)^2$$
$$= \varepsilon(1 - \varepsilon) - \beta(1 - \alpha), \quad (D.7)$$

which is Eq.(8.15).

Since $0 < \beta < \alpha < 1$ and $0 \leq \varepsilon \leq 1$, it follows that

$$c < \varepsilon(1-\varepsilon) = \tfrac{1}{4} - (\tfrac{1}{2}-\varepsilon)^2 \leq \tfrac{1}{4}$$
$$c \geq -\beta(1-\alpha) > -\alpha(1-\alpha) = (\alpha - \tfrac{1}{2})^2 - \tfrac{1}{4} \geq -\tfrac{1}{4}.$$

So we have shown that

$$-\tfrac{1}{4} < c < \tfrac{1}{4}. \tag{D.8}$$

A fixed point of the iteration

$$q_n = c + q_{n+1}^2, \tag{D.9}$$

is

$$q_* = \frac{c}{\tfrac{1}{2} + \sqrt{\tfrac{1}{4} - c}},$$

as can be readily verified by substitution, and to find the domain in which this fixed point is attracting, we define $s_n = q_n - q_*$; and, rewriting $q_n = c + q_{n+1}^2$ in terms of s_n, we have

$$s_n = s_{n+1}\left[1 - \sqrt{1-4c} + s_{n+1}\right]. \tag{D.10}$$

On condition that

$$\left|1 - \sqrt{1-4c}\right| < 1, \tag{D.11}$$

and s_{n+1} is very small, we conclude that q_* is attracting. Indeed, since

$$s_n - s_{n+1} = (s_{n+1} - s_{n+2})\left[1 - \sqrt{1-4c} + s_{n+1} + s_{n+2}\right],$$

the mapping (D.10) is a contraction if $|s_n| \leq \gamma$ and $\left|1 - \sqrt{1-4c} + 2\gamma\right| < 1$. This implies that $\gamma < \sqrt{\tfrac{1}{4}-c}$ when $0 \leq c < \tfrac{1}{4}$, and $\gamma < 1 - \sqrt{\tfrac{1}{4}-c}$ when $-\tfrac{3}{4} < c < 0$. Hence if $|s_N| \leq \gamma$ for very large N, and γ satisfies the above contraction constraint, the iteration backwards to s_0 will be attracted to zero, that is to say q_0 will be attracted to q_*. The domain of attraction of the fixed point is $-\tfrac{3}{4} < c < \tfrac{1}{4}$, and this covers the interval (D.8).

Going back to the original form (D.4) of the iteration, we find that the solution q_* corresponds to the fixed point

$$p_* = \frac{\beta}{\beta + \tfrac{1}{2} - \varepsilon + \sqrt{\beta(1-\alpha) + (\varepsilon - \tfrac{1}{2})^2}}$$

$$= \frac{\beta + \tfrac{1}{2} - \varepsilon - \sqrt{\beta(1-\alpha) + (\varepsilon - \tfrac{1}{2})^2}}{\alpha - 2\varepsilon}. \tag{D.12}$$

D.2 Quadratic Iteration

In the limit that β tends to zero this becomes

$$p_* = \frac{|\frac{1}{2} - \varepsilon| - (\frac{1}{2} - \varepsilon)}{2\varepsilon - \alpha},$$

which is zero if $\varepsilon \leq \frac{1}{2}$. However, if $\varepsilon > \frac{1}{2}$ we find the nontrivial value

$$p_* = \frac{2\varepsilon - 1}{2\varepsilon - \alpha}. \tag{D.13}$$

References

Aikin, Scott F. 2011. *Epistemology and the regress problem.* New York / Oxford: Routledge.
Aikin, Scott F. and Jeanne Peijnenburg. 2014. The regress problem: metatheory, development, and criticism. *Metaphilosophy* 45:139–145.
Alston, William P. 1986. Concepts of epistemic justification. In *Empirical knowledge*, ed. Paul Moser, 23–54. Totowa, New Jersey: Rowman and Littlefield.
Alston, William P. 1989. *Epistemic justification. Essays in the theory of knowledge.* Ithaca / London: Cornell University Press.
Alston, William P. 1993. Epistemic desiderata. *Philosophy and Phenomenological Research* 53:527–551.
Alston, William P. 2005a. *Beyond 'justification': dimensions of epistemic evaluation.* Ithaca / London: Cornell University Press.
Alston, William P. 2005b. Response to Goldman. In *Perspectives on the philosophy of William P. Alston*, eds. Heather D. Battaly and Michael P. Lynch, 137–141. Lanham: Rowman and Littlefield.
Aristotle. 1984a. *Posterior Analytics.* In *The complete works of Aristotle. The revised Oxford translation*, volume 1, ed. Jonathan Barnes, 114–166. Princeton: Princeton University Press.
Aristotle. 1984b. *Physics.* In *The complete works of Aristotle. The revised Oxford translation*, volume 1, ed. Jonathan Barnes, 315–446. Princeton: Princeton University Press.
Aristotle. 1984c. *Metaphysics.* In *The complete works of Aristotle. The revised Oxford translation*, volume 2, ed. Jonathan Barnes, 1552–1728. Princeton: Princeton University Press.
Armstrong, David. 1973. *Belief, truth and knowledge.* Cambridge: Cambridge University Press.
Atkinson, David. 2012. Confirmation and justification: a commentary on Shogenji's measure. *Synthese* 184:49–61.
Atkinson, David and Jeanne Peijnenburg. 2006. Probability without certainty. *Studies in History and Philosophy of Science* 37:442–453.

Atkinson, David and Jeanne Peijnenburg. 2009. Justification by an infinity of conditional probabilities. *Notre Dame Journal of Formal Logic* 50:183–193.
Atkinson, David and Jeanne Peijnenburg. 2010a. Justification by infinite loops. *Notre Dame Journal of Formal Logic* 51:407–416.
Atkinson, David and Jeanne Peijnenburg. 2010b. The solvability of probabilistic regresses. A reply to Frederik Herzberg. *Studia Logica* 94:347–353.
Atkinson, David and Jeanne Peijnenburg. 2010c. Crosswords and coherentism. *Review of Metaphysics* 63:807–820.
Atkinson, David and Jeanne Peijnenburg. 2012. Fractal patterns in reasoning. *Notre Dame Journal of Formal Logic* 53:15–26.
Atkinson, David and Jeanne Peijnenburg. 2013. A consistent set of infinite-order probabilities. *International Journal of Approximate Reasoning* 54:1351–1360.
Atkinson, David, Jeanne Peijnenburg and Theo A. F. Kuipers. 2009. How to confirm the conjunction of disconfirmed hypotheses. *Philosophy of Science* 76:1–21.
Audi, Robert. 1993. *The structure of justification*. Cambridge: Cambridge University Press.
Audi, Robert. 1998. *Epistemology. A contemporary introduction to the theory of knowledge*. London / New York: Routledge.
Aune, Bruce. 1972. Remarks on argument by Chisholm. *Philosophical Studies* 23:327–343.
Bergmann, Michael. 2007. Is Klein an infinitist about doxastic justification? *Philosophical Studies* 134:19–29.
Bergmann, Michael. 2014. Klein and the regress argument. In *Ad infinitum. New essays on epistemological infinitism*, eds. John Turri and Peter D. Klein, 37–54. Oxford: Oxford University Press.
Berker, Selim. 2015. Coherentism via graphs. *Philosophical Issues* 25:322–352.
Beth, Evert W. 1959. *The foundations of mathematics*. Amsterdam: North-Holland. Second revised edition 1968.
Bewersdorf, Benjamin. 2015. Fading foundations and their epistemological consequences. ('Fading foundations en hun epistemologische consequenties'). *Algemeen Nederlands Tijdschrift voor Wijsbegeerte* 107:173–177.
Bird, Alexander. 2007. Justified judging. *Philosophy and Phenomenological Research* 74:81–110.
Black, Oliver. 1988. Infinite regresses of justification. *International Philosophical Quarterly* 28:412–437.
Black, Oliver. 1996. Legal validity and the infinite regress. *Law and Philosophy* 15:339–368.
Bonjour, Laurence. 1976. The coherence theory of empirical knowledge. *Philosophical Studies* 30:281–312.
Bonjour, Laurence. 1985. *The structure of empirical knowledge*. Cambridge, Mass. / London: Harvard University Press.
Booth, Anthony R. 2011. The theory of epistemic justification and the theory of knowledge: a divorce. *Erkenntnis* 75:37–43.
Bovens, Luc and Stephan Hartmann. 2003. *Bayesian epistemology*. Oxford: Clarendon Press.

References

Brown, Patterson. 1966. Infinite causal regression. *Philosophical Review* 75:510–525.
Carnap, Rudolf. 1928. *Der logische Aufbau der Welt.* English translation *The logical structure of the world.* Berkeley / Los Angeles: University of California Press, 1967.
Carnap, Rudolf. 1952. *The continuum of inductive methods.* Chicago: The University of Chicago Press.
Carnap, Rudolf. 1962. *Logical foundations of probability.* Second edition. Chicago: University of Chicago Press.
Carnap, Rudolf. 1980. A basic system of inductive logic, Part II. In *Studies in inductive logic and probability*, ed. Richard C. Jeffrey, Berkeley: University of California Press.
Carroll, Lewis. 1895. What the tortoise said to Achilles. *Mind* 4:278–280.
Chisholm, Roderick M. 1966. *Theory of knowledge.* Englewood Cliffs, New Jersey: Prentice Hall.
Clark, Romane. 1988. Vicious infinite regress arguments. *Philosophical Perspectives – Epistemology* 2:369–380.
Cling, Andrew D. 2004. The trouble with infinitism. *Synthese* 138:101–123.
Cling, Andrew D. 2008. The epistemic regress problem. *Philosophical Studies* 140:401–421.
Cling, Andrew D. 2014. Reasons require reasons. In *Ad Infinitum. New essays on epistemological infinitism*, eds. John Turri and Peter D. Klein, 55–74. Oxford: Oxford University Press.
Cohen, Laurence J. 1977. *The probable and the provable.* Oxford: Oxford University Press.
Comesana, Juan. 2010. Evidentialist reliabilism. *Noûs* 94:571–601.
Conee, Earl, and Richard Feldman. 2004. *Evidentialism.* Oxford: Oxford University Press.
Cornman, James. 1977. Foundational versus nonfoundational theories of empirical justification. *American Philosophical Quarterly* 17:287–297.
Cortens, Andrew. 2002. Foundationalism and the regress argument. *Disputatio* 12:22–37.
Crumley, Jack S. II. 2009. *An introduction to epistemology.* Ontario / New York: Broadview Press.
Crupi, Vincenzo, Branden Fitelson and Katja Tentori. 2007. Probability, confirmation, and the conjunction fallacy. *Thinking and Reasoning* 14:182–199.
Dancy, Jonathan. 1985. *An introduction to contemporary epistemology.* Oxford: Blackwell.
Davidson, Donald. 1963. Action, reasons, and causes. *The Journal of Philosophy* 60:685–700.
Davidson, Donald. 1970. Mental events. In *Experience and theory*, eds. L. Foster and J.W. Swanson. Amhurst: The University of Manchester Press, and London: Duckworth. Reprinted in Davidson 1980, 207–227.
Davidson, Donald. 1980. *Essays on actions and events.* Oxford: Clarendon Press.

Davidson, Donald. 1986. A coherence theory of knowledge and truth. In *Truth and interpretation*, ed. Ernest LePore, 307–319. Oxford: Blackwell.

De Finetti, Bruno. 1972. *Probability, induction and statistics: the art of guessing*. London: Wiley.

De Finetti, Bruno. 1974/1990. *Theory of probability*. Volume 1. Chichester: Wiley Classics Library, John Wiley and Sons.

DeWitt, Richard. 1985. Hume's probability argument of I, iv, 1. *Hume Studies* 11:125–140.

Dieks, Dennis. 2015. Justifiying without foundation ('Rechtvaardigen zonder fundering'). *Algemeen Nederlands Tijdschrift voor Wijsbegeerte* 107:161–165.

Dietrich, Franz and Luca Moretti. 2005. On coherent sets and the transmission of confirmation. *Philosophy of Science* 72:403–424.

Domotor, Zoltan. 1981. Higher order probabilities. *Philosophical Studies* 40:31–46.

Dretske, Fred. 1970. Epistemic operators. *The Journal of Philosophy* 67:1007–1023.

Dretske, Fred. 1971. Conclusive reasons. *Australasian Journal of Philosophy* 49:1–22.

Engel, Mylan Jr. 2014. Positism: the unexplored solution to the epistemic regress problem. *Metaphilosophy* 45:146–160.

Engelsma, Coos. 2014. On Peter Klein's concept of arbitrariness. *Metaphilosophy* 45:192–200.

Engelsma, Coos. 2015. Arbitrary foundations? On Klein's objection to foundationalism. *Acta Analytica* 30:389–405.

Evans, Gareth. 1973. The causal theory of names. *Proceedings of the Aristotelian Society*, Supplementary volume 47:187–225.

Fantl, Jeremy. 2003. Modest infinitsm *Canadian Journal of Philosophy* 33:537–562.

Feldman, Richard, and Earl Conee. 1985. Evidentialism. *Philosophical Studies* 48:15–34.

Firth, Roderick. 1978. Are epistemic concepts reducible to ethical concepts? In *Values and morals: essays in honor of William Frankena, Charles Stevenson, and Richard Brandt*, eds. Alvin I. Goldman and Jaegwon Kim, 215–229. Dordrecht: Reidel.

Fitelson, Branden. 1999. The plurality of Bayesian measures of confirmation and the problem of measure sensitivity. *Philosophy of Science* 66: S363–S378.

Fitelson, Branden. 2003. A probabilistic measure of coherence. *Analysis* 63:194–199.

Fitelson, Branden. 2013. Contrastive Bayesianism. In *Contrastivism in philosophy. New perspectives*, ed. Martijn Blaauw, 64–84. New York / Oxford: Routledge.

Foley, Richard. 1978. Inferential justification and the infinite regress. *American Philosophical Quarterly* 15:311–316.

Foley, Richard. 2012. *When is true belief knowledge?* Oxford / Princeton: Princeton University Press.

Fumerton, Richard. 1995. *Metaepistemology and skepticism*. Boston / London: Rowman and Littlefield.

References

Fumerton, Richard. 2001. Classical foundationalism. In *Resurrecting old-fashioned foundationalism*, ed. Michael R. DePaul, 3–20. Lanham / Oxford: Rowman and Littlefield.

Fumerton, Richard. 2002. Theories of justification. In *The Oxford handbook of epistemology*, ed. Paul K. Moser, 204–233. Oxford: Oxford University Press.

Fumerton, Richard. 2004. Epistemic probability. *Philosophical Issues* 14:149–164.

Fumerton, Richard. 2006. *Epistemology*. Malden, Mass.: Blackwell.

Fumerton, Richard. 2014. Infinitism. In *Ad Infinitum. New essays on epistemological infinitism*, eds. John Turri and Peter D. Klein, 75–86. Oxford: Oxford University Press.

Fumerton, Richard and Ali Hasan. 2010. Foundationalist theories of epistemic justification. *Stanford encyclopedia of philosophy*. http://plato.stanford.edu/entries/justep-foundational

Gaifmann, Haim. 1988. A Theory of Higher Order Probabilities. In *Causation, Chance and Credence*, eds. B. Skyrms and W.L. Harper, 191–219. London: Kluwer.

Gijsbers, Victor. 2015. Fading foundations and the conditional a priori ('Fading Foundations en het conditioneel a priori'). *Algemeen Nederlands Tijdschrift voor Wijsbegeerte* 107:179–183.

Gillett, Carl. 2003. Infinitism *redux*? A response to Klein. *Philosophy and Phenomenological Research* 66:709–717.

Ginet, Carl. 2005. Infinitism is not the solution to the regress problem. In *Contemporary debates in epistemology*, eds. Matthias Steup and Ernest Sosa, 140–149. Oxford: Blackwell.

Goldman, Alvin I. 1967. A causal theory of knowing. *The Journal of Philosophy* 64:357–372.

Goldman, Alvin I. 1986. *Epistemology and cognition*. Cambridge, Mass. / London: Harvard University Press.

Goldman, Alvin I. 2011. Toward a synthesis of reliabilism and evidentialism? Or: Evidentialism's troubles, reliabilism's rescue package. In *Evidentialism and its discontents*, ed. Trent Dougherty, 254–279. Oxford: Oxford University Press.

Gratton, Claude. 2009. *Infinite regress arguments*. Dordrecht: Springer.

Gwiazda, Jeremy. 2011. Infinitism, completability, and computability: reply to Peijnenburg. *Mind* 118:1123–1124.

Haack, Susan. 1993. *Evidence and inquiry: towards reconstruction in epistemology*. Oxford: Blackwell.

Hájek, Alan. 2011. Conditional probability. In *Philosophy of Statistics*, eds. Prasanta S. Bandyopadhyay and Malcolm R. Forster, 95–135. Volume 7 of *Handbook of the Philosophy of Science*, general eds. Dov M. Gabbay, Paul Thagard and John Woods. Amsterdam: Elsevier.

Hájek, Alan and Ned Hall. 1994. The hypothesis of conditional construal of conditional probability. In *Probability and conditionals: belief revision and rational decision*, eds. E. Eells and B. Skyrms, 75–113. Cambridge: Cambridge University Press.

Hankinson, Robert J. 1995. *The sceptics*. London / New York: Routledge.

Harker, Jay E. 1984. Can there be an infinite regress of justified beliefs? *Australasian Journal of Philosophy* 62:255–264.

Hawthorne, James. 2014. Inductive logic. In *Stanford encyclopedia of philosophy*. http://plato.stanford.edu/entries/logic-inductive

Herzberg, Frederik. 2010. The consistency of probabilistic regresses. A reply to Jeanne Peijnenburg and David Atkinson. *Studia Logica* 94:331–345.

Herzberg, Frederik. 2013. The consistency of probabilistic regresses. Some implications for epistemological infinitism. *Erkenntnis* 78:371–382.

Herzberg, Frederik. 2014. The dialectics of infinitism and coherentism. Inferential justification versus holism and coherence. *Synthese* 191:701–723.

Hesslow, Germund. 1976. Discussion: Two notes on the probabilistic approach to causality. *Philosophy of Science* 43:290–292.

Hitchcock, Christopher. 2012. Probabilistic causation. *Stanford encyclopedia of philosophy*. http://plato.stanford.edu/entries/causation-probabilistic

Howson, Colin and Peter Urbach. 1993. *Scientific reasoning. The Bayesian Approach*. 2nd edition. Chicago: Open Court.

Huemer, Michael. 2014. Virtue and vice among the infinite. In *Ad Infinitum. New essays on epistemological infinitism*, eds. John Turri and Peter D. Klein, 87–104. Oxford: Oxford University Press.

Huemer, Michael. 2016. *Approaching infinity*. Basingstoke / New York: Palgrave Macmillan.

Hume, David. 1738/1961. *A Treatise of human nature*. Volume 1. Introduction by A.D. Lindsay. London: J.M. Dent.

Jacobson, Stephen. 1992. Internalism in epistemology and the internalist regress. *Australasian Journal of Philosophy* 70:415–424.

Jäsche, Gottlob B. (Hrsg.)1869/1800. *Immanuel Kant's Logik. Ein Handbuch zu Vorlesungen*. Erläutert von J.H. von Kirchmann. (Dreiundzwanzigster Band *Philosophische Bibliothek oder Sammlung der Hauptwerke der Philosophie alter und neuer Zeit*.) Berlin: L. Heimann.

Jeffreys, Harold. 1939/1961. *Theory of probability*. Oxford: Oxford University Press.

Jenkins-Ichikawa, Jonathan. 2014. Justification as potential knowledge. *Canadian Journal of Philosophy* 44:184–206.

Johnson, William E. 1921. *Logic*. Cambridge: Cambridge University Press.

Johnstone, Henry W. 1996. The rejection of infinite postponement as a philosophical argument. *Journal of Speculative Philosophy* 10:92–104.

Kajamies, Timo. 2009. A quintet, a quartet, a trio, a duo? The epistemic regress problem, evidential support, and skepticism. *Philosophia* 37:525–535.

Keynes, John M. 1921 *A treatise on probability*. London: Macmillan.

Klein, Peter. 1998. Foundationalism and the infinite regress of reasons. *Philosophy and Phenomenological Research* 58:919–925.

Klein, Peter. 1999. Human knowledge and the infinite regress of reasons. *Philosophical Perspectives – Epistemology* 13:297–325.

Klein, Peter. 2003. When infinite regresses are not vicious. *Philosophy and Phenomenological Research* 66:718–729.

References

Klein, Peter. 2005. Infinitism is the solution to the regress problem. In *Contemporary debates in epistemology*, eds. Matthias Steup and Ernest Sosa, 131–140. Oxford: Blackwell.

Klein, Peter. 2007a. Human knowledge and the infinite progress of reasoning. *Philosophical Studies* 134:1–17.

Klein, Peter. 2007b. How to be an infinitist about doxastic justification. *Philosophical Studies* 134:25–29.

Klein, Peter. 2010. Why not infinitism? In *Proceedings of the Twentieth World Congress of Philosophy*, Volume 5, ed. Richard Cobb-Stevens, 199–208. Charlottesville, Virginia: Philosophy Documentation Center.

Klein, Peter. 2011a. Infinitism. In *The Routledge companion to epistemology*, eds. Sven Bernecker and Duncan Pritchard, 245–256. New York: Routledge.

Klein, Peter. 2011b. Infinitism and the epistemic regress problem. In *Conceptions of knowledge*, ed. Stefan Tolksdorf, 487–508. Berlin: Walter de Gruyter.

Klein, Peter. 2014. Reasons, reasoning, and knowledge: a proposed rapprochement between infinitism and foundationalism. In *Ad infinitum. New essays on epistemological infinitism,* eds. John Turri and Peter D. Klein, 105–124. Oxford: Oxford University Press.

Kolmogorov, Andrey N. 1933. *Grundbegriffe de Wahrscheinlichkeitsrechnung*. Berlin: Springer.

Kripke, Saul A. 1970/1980. *Naming and necessity*. Cambridge, Mass.: Harvard University Press. Second printing 1981.

Kvanvig, Jonathan L. 2014. Infinitist justification and proper basing. In *Ad Infinitum. New essays on epistemological infinitism*, eds. John Turri and Peter D. Klein, 125–142. Oxford: Oxford University Press.

Kyburg, Henry E. 1987. *Higher order probabilities and intervals.* Volume 36 of Technical Report Department of Computer Science, University of Rochester.

Legum, Richard A. 1980. Probability and foundationalism: another look at the Lewis-Reichenbach debate. *Philosophical Studies* 38:419–425.

Lehrer, Keith. 1974. *Knowledge*. Oxford: Clarendon Press.

Lehrer, Keith, 1981. The evaluation of method: a hierarchy of probabilities among probabilities. *Grazer Philosophische Studien* 12-13:131–141.

Lehrer, Keith. 1997. Justification, coherence and knowledge. *Erkenntnis* 50:243–258.

Leite, Adam. 2005. A localist solution to the regress of justification. *Australasian Journal of Philosophy* 83:395–421.

Lemos, Noah. 2007. *An introduction to the theory of knowledge*. Cambridge: Cambridge University Press.

Leplin, Jarrett. 2009. *A theory of epistemic justification*. Philosophical Studies Series 112. Dordrecht: Springer.

Lewis, Clarence I. 1929. *Mind and the world-order. An outline of a theory of knowledge*. New York: C. Scribner's Sons.

Lewis, Clarence I. 1946. *An analysis of knowledge and valuation.* La Salle, Illinois: Open Court.

Lewis, Clarence I. 1952. The given element in empirical knowledge. *The Philosophical Review* 61:168–172.

Lewis, David. 1980. A subjectivist's guide to objective chance. In *Studies in Inductive Logic and Probability. Vol II*, ed. R. C. Jeffrey, 263–293. Berkeley: University of California Press.

Littlejohn, Clayton. 2012. *Justification and the truth-connection.* Cambridge: Cambridge University Press.

Mandelbrot, Benoît B. 1977. *The fractal geometry of nature.* Second printing with update, 1982. New York: W.H. Freeman and Co.

McGrew, Lydia and Timothy McGrew. 2000. Foundationalism, transitivity and confirmation. *The Journal of Philosophical Research* 25:47–66.

McGrew, Lydia and Timothy McGrew. 2008. Foundationalism, probability, and mutual support. *Erkenntnis* 68:55–77.

Miller, David. 1966. A paradox of information. *British Journal for the Philosophy of Science*, 17:59–61.

Moretti, Luca. 2007. Ways in which coherence is confirmation conducive. *Synthese* 157:309–319.

Moser, Paul K. 1984. A defense of epistemic intuitionism. *Metaphilosophy* 15:196–209.

Moser, Paul K. 1985. Whither infinite regresses of justification. *Southern Journal of Philosophy* 23:65–74.

Moser, Paul K., Dwayne H. Mulder, and J.D. Trout. 1998. *The theory of knowledge. A thematic introduction.* New York / Oxford: Oxford University Press.

Nelkin, Dana K. 2000. The lottery paradox, knowledge and rationality. *The Philosophical Review* 109:373–409.

Nesson, Charles R. 1979. Reasonable doubt and permissive inferences: the value of complexity. *Harvard Law Review* 92:1187–1225.

Neta, Ram. 2014. Klein's case for infinitism. In *Ad Infinitum. New essays on epistemological infinitism*, eds. John Turri and Peter D. Klein, 143–161. Oxford: Oxford University Press.

Neurath, Otto. 1932-1933. Protocol sentences. *Erkenntnis* 3. Reprinted in *Logical positivism*, ed. A.J. Ayer, 199–208, Toronto: The Free Press, 1959. First paperback edition 1966. Translation from the German by George Schick.

Nolan, Daniel. 2001. What's wrong with infinite regresses? *Metaphilosophy* 32:523–538.

Norman, Andrew. 1997. Regress and the doctrine of epistemic original sin. *The Philosophical Quarterly* 47:477–494.

Nozick, Robert. 1981. *Philosophical explanations.* Cambridge, Mass.: Harvard University Press.

Oakley, I.T. (Tim) 1976. An argument for scepticism concerning justified beliefs. *American Philosophical Quarterly* 13:221–228.

Olsson, Erik J. 2001. Why coherence is not truth conducive *Analysis* 61:236–241.

Olsson, Erik J. 2002. What is the problem of coherence and truth? *The Journal of Philosophy* 99:246–272.

Olsson, Erik J. 2005a. *Against coherence: truth, probability, and justification.* Oxford: Oxford University Press.
Olsson, Erik J. 2005b. The impossibility of coherence. *Erkenntnis* 63:387–412.
Owens, Joseph. 1962. Aquinas on infinite regress. *Mind* 71:244–246.
Pastin, Mark. 1975. C.I. Lewis's radical foundationalism. *Noûs* 9:407–420.
Pearl, Judea. 2000. *Causality.* Cambridge: Cambridge University Press.
Peijnenburg, Jeanne. 1998. The dual character of causality. *Dialektik* 2:71–81.
Peijnenburg, Jeanne. 2007. Infinitism regained. *Mind* 116:597–602.
Peijnenburg, Jeanne. 2010. Ineffectual foundations: reply to Gwiazda. *Mind* 119:1125–1133.
Peijnenburg, Jeanne. 2015. Reply ('Repliek'). *Algemeen Nederlands Tijdschrift voor Wijsbegeerte* 107:199–211.
Peijnenburg, Jeanne and David Atkinson. 2008. Probabilistic justification and the regress problem. *Studia Logica* 89:333–341.
Peijnenburg, Jeanne and David Atkinson. 2011. Grounds and limits. Reichenbach and foundationalist epistemology. *Synthese* 81:113–124.
Peijnenburg, Jeanne and David Atkinson. 2014a. Can an infinite regress justify anything? In *Ad Infinitum. New essays on epistemological infinitism*, eds. John Turri and Peter D. Klein, 162–179. Oxford: Oxford University Press.
Peijnenburg, Jeanne and David Atkinson. 2014b. The need for justification. *Metaphilosophy* 45:201–210.
Peijnenburg, Jeanne and Sylvia Wenmackers. 2014. Infinite regress in decision theory, philosophy of science, and formal epistemology. *Synthese* 191:627–628.
Peirce, Charles S. 1868. Questions concerning certain faculties claimed for man. *The Journal of Speculative Philosophy* 2, 103–114.
Plantinga, Alvin. 1993. *Warrant: the current debate.* Oxford: Oxford University Press.
Podlaskowski, Adam C. and Joshua A. Smith. 2011. Infinitism and epistemic normativity. *Synthese* 178:515–527.
Podlaskowski, Adam C. and Joshua A. Smith. 2014. Probabilistic regresses and the availability problem for infinitism. *Metaphilosophy* 45:211–220.
Pollock, John. 1974. *Knowledge and justification.* Princeton: Princeton University Press.
Popper, Karl R. 1959. *The logic of scientific discovery.* New York: Basic Books.
Post, John. 1980. Infinite regresses of justification and explanation. *Philosophical Studies* 38:31–52.
Post, John and Derek Turner. 2000. Sic transitivity: reply to McGrew and McGrew. *The Journal of Philosophical Research* 25:67–82.
Poston, Ted. 2012. Basic reasons and first philosophy: a coherentist view of reasons. *The Southern Journal of Philosophy* 50:75–93.
Poston, Ted. 2014. Finite reasons without foundations. *Metaphilosophy* 45:182–191.
Pritchard, Duncan. 2005. *Epistemic luck.* Oxford: Oxford University Press.
Pritchard, Duncan. 2006. *What is this thing called knowledge?* Second edition 2010. New York / Abingdon: Routledge.

Pritchard, Duncan. 2007. Anti-luck epistemology. *Synthese* 158:277–298.

Pritchard, Duncan. 2008. Knowledge, luck, and lotteries. In *New waves in epistemology,* eds. Vincent Hendricks and Duncan Pritchard, 28–51. London: Palgrave Macmillan.

Quine, Willard V.O. and J.S. Ullian. 1970. *The web of belief.* New York: Random House.

Quine, Willard V.O. 2008. Lectures on David Hume's philosophy: 1946. Edited by James G. Buickerood in: W.V. Quine, *Confessions of a confirmed externalist and other essays,* eds. D. Follesdal and D.B. Quine, Cambridge, Mass.: Harvard University Press. New York: Random House.

Reichenbach, Hans. 1938. *Experience and prediction.* Chicago: University of Chicago Press.

Reichenbach, Hans. 1952. Are phenomenological reports absolutely certain? *The Philosophical Review* 61:147–159.

Reichenbach, Hans. 1956. *The direction of time.* Berkeley: University of California Press.

Reichenbach, Maria and Robert Cohen (eds.). 1978. *Hans Reichenbach. Selected writings 1909-1953.* Dordrecht: Reidel.

Rényi, Alfred. 1970/1998 *Foundations of probability.* Mineola, New York: Dover Publications.

Rescher, Nicholas. 1973. *The coherence theory of truth.* Oxford: Oxford University Press.

Rescher, Nicholas. 2005. Issues of Infinite Regress. In: N. Rescher, *Scholastic Meditations,* 92–109. Washington: The Catholic University of America Press.

Rescher, Nicholas. 2010. *Infinite regress. The theory and history of varieties of change.* New Brunswick: Transaction Publishers.

Rescorla, Michael. 2009. Epistemic and dialectical regress. *Australasian Journal of Philosophy* 87:43–60.

Rescorla, Michael. 2014. Can perception halt the regress of justifications? In *Ad infinitum. New essays on epistemological infinitism,* eds. John Turri and Peter D. Klein, 179–200. Oxford: Oxford University Press.

Reynolds, Steven L. 2013. Justification as the appearance of knowledge. *Philosophical Studies* 163:367–383.

Roche, William A. 2012. A reply to Cling's 'The epistemic regress problem'. *Philosophical Studies* 159:263–276.

Roche, William A. and Tomoji Shogenji. 2014. Dwindling confirmation. *Philosophy of Science* 81:114–137.

Roche, William A. 2016. Foundationalism with infinite regresses of probabilistic support? Manuscript.

Rorty, Richard. 1982. *Consequences of pragmatism.* Brighton: The Harvester Press.

Russell, Bertrand. 1906. On the nature of truth. *Proceedings of the Aristotelian Society* 7:28–49.

Russell, Bertrand. 1948. *Human knowledge.* London: George Allen and Unwin.

Sanford, David H. 1975. Infinity and vagueness. *The Philosophical Review* 84:520–535.

References

Sanford, David H. 1984. Infinite regress arguments. In *Principles of philosophical reasoning*, ed. James H. Fetzner, 93–117. Totowa, New Jersey: Roman and Allanheld.

Savage, Leonard J. 1954/1972. *The foundations of statistics*. New York: Dover. Second revised edition.

Sellars, Wilfrid. 1956. Empiricism and the philosophy of mind. In *The foundations of science and the concepts of psychology and psychoanalysis*. Volume 1 of *Minnesota Studies in the Philosophy of Science*, eds. Herbert Feigl and Michael Scriven, 253–329. Minneapolis: University of Minnesota Press.

Sextus Empiricus. 1994. *Outlines of scepticism*. Translated by Julia Annas and Jonathan Barnes. Cambridge: Cambridge University Press.

Shogenji, Tomoji. 1999. Is coherence truth conducive? *Analysis* 59:338–345.

Shogenji, Tomoji. 2003. A condition for transitivity in probabilistic support. *British Journal for the Philosophy of Science* 54:613–616.

Shogenji, Tomoji. 2012. The degree of epistemic justification and the conjunction fallacy. *Synthese* 184:29–48.

Simson, Rosalind S. 1986. An internalist view of the epistemic regress problem. *Philosophy and Phenomenological Research* 48:179–208.

Skyrms, Brian. 1980. Higher order degrees of belief. In *Prospects for pragmatism: essays in memory of F.P. Ramsey*, ed. D.H. Mellor, 109–137. Cambridge: Cambridge University Press.

Smith, Martin. 2010. What else justification could be. *Noûs* 44:10–31.

Smith, Martin. 2016. *Between probability and certainty*. Oxford: Oxford University Press.

Sosa, Ernest. 1979. Epistemic presupposition. In *Justification and knowledge. New studies in epistemology*, ed. George S. Pappas, 79–92. Dordrecht / Boston / London: D. Reidel Publishing Company.

Sosa, Ernest. 1980. The raft and the pyramid: coherence vs foundationalism in the theory of knowledge. *Midwest Studies in Philosophy*. 5:3–25.

Sosa, Ernest. 1999a. How to defeat opposition to Moore. *Philosophical Perspectives* 13:141–154.

Sosa, Ernest. 1999b. How must knowledge be modally related to what is known? *Philosophical Topics* 26:373–384.

Sosa, Ernest. 2014. Infinitism. In *Ad infinitum. New essays on epistemological infinitism*, eds. John Turri and Peter D. Klein, 201–209. Oxford: Oxford University Press.

Spirtes, Peter, Clark N. Glymour, and Richard Scheines. 1993. *Causation, prediction and search*. New York: Springer Verlag.

Spohn, Wolfgang. 2012. *The laws of belief: ranking theory and its philosophical applications*. Oxford: Oxford University Press.

Steup, Matthias. 1989. The regress of metajustification. *Philosophical Studies* 55:41–56.

Steup, Matthias. 2005. Epistemology. *Stanford encyclopedia of philosophy*. http://plato.stanford.edu/entries/epistemology

Stoutland, Frederick M. !970. The logical connection argument. *American Philosophical Quarterly* 7:117–129.
Sutton, Jonathan. 2007. *Without justification*. Boston: M.I.T. Press.
Swinburne, Richard. 2001. *Epistemic justification*. Oxford: Oxford University Press.
Turri, John. 2010. On the relationship between propositional and doxastic justification. *Philosophy and Phenomenological Research* 80:312–326.
Turri, John. 2009. On the regress argument for infinitism. *Synthese* 166:157–163.
Turri, John and Peter D. Klein (eds.) 2014. *Ad Infinitum. New essays on epistemological infinitism*. Oxford: Oxford University Press.
Uchii, Soshichi. 1973. Higher order probabilities and coherence. *Philosophy of Science* 40:373–381.
Vahid, Hamid. 2011. Externalism/internalism. In *The Routledge companion to epistemology*, eds. Sven Bernecker and Duncan Pritchard, 144–155. New York: Routledge.
Van Benthem, Johan. 2014. *Fanning the flames of reason*. Valedictory lecture delivered on the occasion of his retirement as University Professor at the University of Amsterdam on 26 September 2014.
Van Benthem, Johan. 2015. Justification as dynamic logical process ('Rechtvaardigen als dynamisch logisch proces'). *Algemeen Nederlands Tijdschrift voor Wijsbegeerte* 107:147–153.
Van Cleve, James. 1977. Probability and certainty. *Philosophical Studies* 32:323–334.
Van Cleve, James. 1992. Supervenience and referential indeterminacy. *The Journal of Philosophy* 89:344–362.
Van Woudenberg, René and Ronald Meester. 2014. Infinite epistemic regresses and internalism. *Metaphilosophy* 45:221–231.
Walker, Ralph C.S. 1997. Theories of truth. In *A companion to the philosophy of language*, eds. Bob Hale and Crispin Wright, 309–330. Oxford / Malden: Blackwell.
Wengert, Robert G. 1971. The logic of essentially ordered causes. *Notre Dame Journal of Formal Logic* 12:406–422.
Wieland, Jan Willem. 2014. *Infinite regress arguments*. Dordrecht: Springer.
Williams, Christopher J.F. 1960. Hic autem non est procedere in infinitum... *Mind* 69:403–405.
Williams, John N. 1981. Justified belief and the infinite regress argument. *American Philosophical Quarterly* 18:85–88.
Williams, Michael. 2004. The Agrippan argument and two forms of skepticism. In *Pyrrhonian skepticism*, ed. Walter Sinnott-Armstrong, 121–145. Oxford: Oxford University Press.
Williams, Michael. 2010. Descartes' transformation of the sceptical tradition. In *Ancient discourse*, ed. Richard Bett, 288–313. Cambridge: Cambridge University Press.
Williams, Michael. 2014. Avoiding the regress. In *Ad infinitum. New essays on epistemological infinitism,* eds. John Turri and Peter D. Klein, 227–242. Oxford: Oxford University Press.

References

Williamson, Timothy. 2000. *Knowledge and its limits*. Oxford: Oxford University Press.

Wright, Stephen. 2013. Does Klein's infinitism offer a response to Agrippa's trilemma? *Synthese* 190:1113–1130.

Zagzebski, Linda. 2014. First person and third person reasons and the regress problem. In *Ad Infinitum. New essays on epistemological infinitism*, eds. John Turri and Peter D. Klein, 243–255. Oxford: Oxford University Press.

Open Access This chapter is licensed under the terms of the Creative Commons Attribution 4.0 International License (http://creativecommons.org/licenses/by/4.0/), which permits use, sharing, adaptation, distribution and reproduction in any medium or format, as long as you give appropriate credit to the original author(s) and the source, provide a link to the Creative Commons license and indicate if changes were made.

The images or other third party material in this chapter are included in the chapter's Creative Commons license, unless indicated otherwise in a credit line to the material. If material is not included in the chapter's Creative Commons license and your intended use is not permitted by statutory regulation or exceeds the permitted use, you will need to obtain permission directly from the copyright holder.

Index

Agrippa's trilemma, 1, 3–5, 7, 9, 10, 16
Aikin, S., viii, 34, 46, 101, 102, 119–121, 124, 130, 131, 173
Alston, W., 14, 27, 29, 33, 36, 52–54, 56, 57, 121, 173
Aquinas, T., 7, 17–21
Aristotle, 5, 6, 9, 11, 16–19, 21, 22, 34, 101, 121
Armstrong, D., 15, 30
Audi, R., 3, 11
Aune, B., 143
Ayer, A., 9

bacteria, 61, 78–82, 84, 96, 126, 184, 185, 188, 214
basing, 52, 56, 57
belief, 1, 2, 5, 7, 9, 10, 12, 13, 16, 17, 26, 27, 29–37, 39–41, 43, 44, 46, 47, 49, 52–55, 57, 78, 83, 86–94, 96–98, 100–103, 108, 109, 115, 116, 120–122, 127, 131, 134, 143, 147, 182, 190, 209

Bergmann, M., 87–89, 102
Berkeley, G., 8
Berker, S., 131, 189, 190
Beth, E., 23
Bewersdorf, B., 97
Bird, A., 28
Black, O., 42
Bonjour, L., 3, 9, 10, 90, 120
Booth, A., 28
Bovens, L., 10
Brown, P., 19–22

Carnap, R., 9, 26, 37, 95, 134, 138
Carroll, L., 151
causal chain, 1, 17, 98, 185–187, 199
causal graphs, 185–190
Chisholm, R., 26, 36, 143
Clark, R., 42, 121, 174
Cling, A., 29, 42, 133
closure, 119, 134, 136–139, 203–205
Cohen, L., 50
Cohen, R., 68
coherentism, 8–12, 60, 83, 89–91, 168, 170, 173, 189, 190

233

Comesana, J., 32
Conee, E., 30, 31, 33
convergence, 67, 72, 75, 89, 99, 109, 149, 152, 155, 164, 172, 180, 181, 191, 194, 195, 200
Cornman, J., 29, 35, 120, 128
Cortens, A., 173
Crumley, J., 16
Crupi, V., 45

Dancy, J., 90, 121, 123
Davidson, D., 9, 10, 31, 44
De Finetti, B., 95, 147
Descartes, R., 7, 8, 36
DeWitt, R., 150
Dieks, D., 69
Dietrich, F., 10
Diogenes Laertius, 3
directed acyclic graphs, 186
Domotor, Z., 147
Dretske, F., 33

Engel, M., 92
Engelsma, C., 13, 50
ethics, 26, 30, 127
Evans, G., 98
evidentialism, 31, 32, 39, 41
evidential support, 46–51
externalism, 25, 30, 39–41

fading foundations, 82–89, 92, 97–99, 105, 109–112, 133, 134, 141, 145, 157, 160, 180, 185, 186, 190, 207, 211
Fantl, J., 92, 93
Feldman, R., 30, 31, 33

finite mind objection, 101–103, 105, 106, 109, 111, 113, 114, 116, 121, 122, 124
Firth, R., 87
Fitelson, B., 10, 38, 45, 134, 138
fixed points, 72, 179, 213, 214, 216
Foley, R., 28, 35, 36, 120
foundationalism, 1, 6–13, 16, 87, 97, 98, 122, 127, 131, 169, 182, 183
fractal, 167, 169, 176–181, 183, 185, 215
Fumerton, R., 16, 17, 29, 36, 37, 42, 43, 62, 102, 103, 122, 123, 127, 168, 182–184

Gaifmann, H., 147
Gijsbers, V., 97
Gillett, C., 123
Ginet, C., 121
Glymour, C., 186
Goldman, A., 28, 30, 32, 40–44, 48
Goodman, N., 48, 63
Gratton, C., 15
Gwiazda, J., 105, 107, 108

Hájek, A., 95
Hankinson, R., 22, 121
Harker, J., 131
Hartmann, S., 10
Hasan, A., 62, 122
Hawthorne, J., 208
Hegel, G., 9, 167, 169
Herzberg, F., 60, 76, 91, 105, 106, 115, 160, 187

INDEX 235

Hesslow, G., 186
Hitchcock, C., 186
Howson, C., 146, 209
Huemer, M., 3, 15
Hume, D., 8, 48, 62, 67, 148, 149

infinitism, 1, 8, 10–14, 16, 17, 27, 28, 87, 92, 103–106, 116, 131, 133
internalism, 25, 30, 39–41

Jäsche, G., 25
Jacobson, S., 40
Jeffreys, H., 95
Jenkins Ichikawa, J., 28
Johnson, W., 95
Johnstone, H., 121, 123
justification
  completion, 116
  doxastic, 40, 53, 54, 79, 83, 86–90, 92, 93, 103, 104, 114, 115
  emergence, 71, 90–93, 98, 99, 105, 110, 133, 134
  epistemic, 2, 7, 12–14, 23, 25–32, 34–39, 41–47, 49–54, 56, 57, 59, 60, 75, 78, 83, 87, 89, 94, 97–100, 102–105, 107–109, 113, 115, 119, 122–125, 128, 132, 133, 138–143, 167, 169, 173, 174, 179–181, 183, 185, 187, 188, 190, 194
  propositional, 44, 53, 56, 57, 79, 83, 86–88, 91, 92, 103, 104, 114

trade-off, 51, 75, 90, 100, 101, 107–114, 133, 193, 195
trees, 173–176, 187–189

Kajamies, T., 42, 121
Kant, I., 8, 9, 25–27, 30, 154, 170, 171
Keynes, J., 95
Klein, P., viii, 11–16, 23, 27, 46, 87–92, 103–105, 116, 120, 130, 131
Kolmogorov, A., 39, 45, 51–53, 57, 60, 66, 94, 95, 98, 140
Kripke, S., 98
Kuipers, T., 138
Kvanvig, J., 87, 104
Kyburg, H., 147

Legum, R., 64
Lehrer, K., 10, 62, 149
Leite, A., 113
Lemos, N., 16, 102
Leplin, J., 46
Lewis, C., 54, 59–72, 74–78, 80, 143, 144, 147, 149, 158, 159
Lewis, D., 34, 48, 147
Lewis-Reichenbach dispute, 61–67, 70–72, 79
Littlejohn, C., 28
Locke, J., 8, 36
lottery paradox, 136

Mandelbrot, B., 167, 169, 177
Mann, T., v, 30
Markov condition, 140, 141, 160, 162, 186–188, 191, 199–201

Markov process, 72
McGrew, L., 42, 90
McGrew, T., 42, 90
Meester, R., 40, 43
Miller, D., 146, 147, 152, 156
Montaigne, M. de, 7
Mooij-Valk, H. and S., 30
Moore, G., 36
Moretti, L., 10
Moser, P., 12, 16, 133
Mulder, D., 16
mushrooming out, 182–184

Nauta, L., 7
Nawar, T., 3
Nelkin, D., 46, 47
Nesson, C., 50
Neta, R., 29
Neurath, O., 9, 91
Nietzsche, F., 9
no starting point, 119, 121–126, 128, 139, 183
Nolan, D., 15
Norman, A., 113
normic support, 45, 47–51
Nozick, R., 33, 34

Oakley, I., 34, 120, 130
Olsson, E., 10
Owens, J., 20

Pastin, M., 65, 66, 79
Pearl, J., 147, 186
Peirce, C., 12, 154
Plantinga, A., 28, 56
Plato, 26, 27
Podlaskowski, A., 105–108, 115–117

Poincaré, H., 169
Pollock, J., 35, 120, 125, 128
Popper, K., 95
Post, J., 29, 35, 42, 120, 128–130
Poston, T., 87, 113
Pritchard, D., 16, 34
probability
　higher-order probabilities, 67, 143–145, 147, 148, 153, 156, 157, 160, 184
　imprecise probabilities, 81
　network, 75, 83, 108, 133, 167–169, 173, 174, 180–183, 187, 195, 213
　probabilistic loops, 4, 9, 167–173
　probabilistic support, 25, 34, 36–43, 45, 46, 50, 51, 53, 55, 59–61, 75, 78, 81, 84, 85, 89, 91, 93, 97, 99, 100, 104, 108–112, 117, 119, 120, 123, 124, 126, 131, 134–136, 139–141, 158, 159, 167, 169, 179, 182, 192, 193, 195, 200, 214
　second-order probabilities, 145–148, 150, 156
　total probability, 54, 57, 66, 68–70, 72, 73, 79, 84, 98, 126, 127, 135, 144, 149, 151, 156, 159, 161, 175, 184, 191, 192, 208
proper name, 98
Pyrrho of Elis, 3–5, 7, 8

Quine, W., 10, 67, 149

INDEX 237

reductio argument, 120, 128, 130–132, 139
regress
  benign, 15, 19, 22, 59, 74–76, 83, 99
  causal, 17–19, 21, 22, 60, 186
  of entailment, 29, 30, 33–35, 60, 61, 77, 99, 104, 119, 123–126, 128–131, 133, 139, 140, 170, 191, 198
  exceptional class, 59, 61, 74–78, 83, 98, 99, 124, 149, 167, 169, 173, 191, 195, 197, 200
  probabilistic, 59–62, 69, 73–80, 83, 84, 86, 99, 105, 108, 115, 116, 119, 120, 124–126, 131–133, 144, 145, 149, 156, 164, 169, 171, 187, 199, 207, 211
  usual class, 59, 61, 74–76, 78, 85, 99, 104, 125, 132, 135, 141, 149, 157, 160, 164, 167, 169, 172, 173, 187, 191, 194–197, 200, 207
  vicious, 14, 15, 17–19, 21, 22, 59, 61, 64, 72, 76, 78, 105, 122, 123, 127, 154
Reichenbach, H., 39, 59–61, 63, 64, 66, 68, 69, 71, 72, 79, 143, 159, 162, 186, 188, 200
Reichenbach, M., 68
reliabilism, 31, 32, 39, 41
Rényi, A., 95
Rescher, N., 10, 15, 22, 62, 82, 90, 95, 105–107, 145, 149–155

Rescorla, M., 12, 107, 113
Reynolds, S., 28
rigid designator, 98
Roche, W., 42, 133, 168, 200
Rorty, R., 26, 27
Russell, B., 9, 10, 54, 66–69, 86, 143, 149

Sanford, D., 82
Savage, L., 148, 150
scepticism, 3–8, 34, 35, 88, 96, 104, 133, 182
Scheines, R., 186
Sellars, W., 167, 168
Sextus Empiricus, 3, 4, 7, 121
Shogenji, T., 10, 46, 134, 137–139, 200, 203
Simson, R., 40
Skyrms, B., 146, 147, 157
Smith, J., 105–108, 115–117
Smith, M., 28, 34, 45–51
Sosa, E., 34, 90, 91, 174
Spirtes, P., 186
Spohn, W., 48, 98
Steup, M., 29, 36, 105
Stoutland, F., 44
Sutton, J., 28
Swinburne, R., 27, 41

Tentori, K., 45
threshold, 32, 38, 39, 46, 50, 119, 132–137, 139, 140
Trout, J., 16
Turri, J., 12, 23, 87, 92

Uchii, S., 147
Ullian, J., 10
Urbach, P., 146, 209

Vahid, H., 30
Van Benthem, J., 69, 107
Van Cleve, J., 12, 63, 64, 72
Van Woudenberg, R., 40, 43
Verhaegh, S., 67

Walker, R., 10
washing out, 86, 207, 209, 211
Wengert, R., 20

Wenmackers, S., viii
Wieland, J., 15
Williams, C., 20
Williams, J., 102
Williams, M., 8, 29, 30, 87
Williamson, T., 28, 34, 46
Wittgenstein, L., 30
Wright, S., 105

Zagzebski, L., 113

The manufacturer's authorised representative in the EU is Springer Nature Customer Service Centre GmbH, Europaplatz 3, 69115 Heidelberg, Germany. If you have any concerns regarding our products, please contact ProductSafety@springernature.com

Printed and bound by CPI Group (UK) Ltd, Croydon, CR0 4YY

26/03/2026

02078916-0004